If Walls Could Speak

Also by Moshe Safdie:

With Intention to Build: The Unrealized Concepts, Ideas, and Dreams of Moshe Safdie

Megascale, Order & Complexity

The City after the Automobile: An Architect's Vision
(with Wendy Kohn)

Jerusalem: The Future of the Past

Beyond Habitat by 20 Years

The Harvard Jerusalem Studio: Urban Designs for the Holy City

Form & Purpose

For Everyone a Garden

Beyond Habitat

If Walls Could Speak
My Life in Architecture

MOSHE SAFDIE

Atlantic Monthly Press

New York

Book design: Michael Gericke, with Reid Parsekian

FIRST EDITION

Published simultaneously in Canada
Printed in Canada

First Grove Atlantic hardcover edition: September 2022

Library of Congress Cataloging-in-Publication data is available for this title.

ISBN 978-0-8021-5832-1 eISBN 978-0-8021-5834-5

Atlantic Monthly Press
an imprint of Grove Atlantic
154 West 14th Street
New York, NY 10011

Distributed by Publishers Group West

groveatlantic.com

22 23 24 25   10 9 8 7 6 5 4 3 2 1

For Michal,
on our fiftieth together

CONTENTS

A swim in the
infinity pool,
Marina Bay Sands,
Singapore, 2010.

A Week in the Life

The offices of Safdie Architects are lodged in a four-story industrial brick structure, built at the end of the nineteenth century. It sits unobtrusively on a side street in Somerville, Massachusetts, fifteen minutes' walk from the Harvard Graduate School of Design. Over the years, the building has served as a wicker factory, a wine-wholesaling warehouse, and a motor-maintenance shop. Our firm occupied the premises in 1982. We cleaned up the interior but kept the bones basically intact. The brick structure is supported by a frame of massive wood beams and columns, with elegant cast-iron capitals—the topmost feature—forming the connection between column and beam. Large, subdivided, steel-framed windows punctuate the brick exterior. Eventually, we expanded the building, adding three bays on the back and a new fourth floor. There was ivy climbing the building, here and there, when we bought the place. Today the exterior is covered completely with a combination of Boston and English ivy. From the outside, the building appears leafy and green, with hardly any brick visible at all. In a gentle breeze, it seems to breathe. A raccoon has established a nest outside the window of my personal office, sheltered by the ivy. It sleeps against the pane most of the day—a calming presence.

On this Thursday morning in early 2020 I arrive at the office by car, a five-minute drive from my house in Cambridge. I had been back in the office for an uninterrupted two-week stretch—something of a rarity. In two days, I will be leaving again. Somerville may be the location of our headquarters, but the practice is worldwide, and much of the work in recent years has come from Asia. The pandemic, when it came, would momentarily disrupt our habits but not the work itself. My associates and I had just been in Singapore for the official opening of Jewel Changi, a retail-and-entertainment complex at the airport centered on a vast, toroidal dome of glass that caps a terraced tropical forest. A cascade of rainwater—the tallest indoor waterfall in the world—falls continuously from the oculus. At the opening ceremony, the prime minister had cut the ribbon. He had generous things to say about the design. Members of my office, along with my family, joined the celebration. We also celebrated the marriage of Jaron Lubin, a partner in our firm, to Helen Han, also an architect, with the waterfall as a backdrop. Jaron had served

as one of two project principals in charge of Jewel. In Singapore, the family and the traveling staff stayed at Marina Bay Sands, the waterfront development that our firm had also designed and built—now a decade old and a Singapore landmark. If you have seen *Crazy Rich Asians*, then you have seen Marina Bay Sands, which is the setting for many scenes in the movie. The opening of the Jewel complex was the culmination of a six-year effort. Hundreds of thousands of people flocked to Jewel on the first day, for the soft opening, and then kept coming. (There would be 50 million visitors in the first six months.) The mood upon our return to the office was jubilant.

The premises in Somerville are a welcoming place on any day, under any circumstance—designed that way to enhance the functional, collegial spirit that a firm such as ours needs, and also to showcase some of our values in a modest way. From the glass front doors, a visitor looks down a passage that opens into a view of the full length of the building, all the way to the soft glow of natural light at the rear. The interior is expansive and layered—catwalks, open staircases, unimpeded views. From any space on any floor, you can see the rest of the activity on the floor, and from various places on any floor you can see parts of every other floor. I sometimes wonder what the artist M. C. Escher might have made of the place.

You enter the building on what is actually the second floor—it seems to a visitor like the ground floor, but there is a level underneath devoted to our business offices and, mostly, to the model workshop, a place where I could happily pass many hours of any day. On the lobby walls hang photographs of some of our signature projects—Jewel and Marina Bay Sands, of course; the Crystal Bridges Museum of American Art, in Arkansas, commissioned by Alice Walton; and Habitat '67, in Montreal, the residential complex designed for Expo 67, which jump-started my career—but also, nearby, photographs of the Holocaust History Museum at Yad Vashem, in Israel; the Skirball Cultural Center, in Los Angeles; the United States Institute of Peace headquarters, in Washington, D.C.; and the Khalsa Heritage Centre, in Punjab, India. There are also many photographs of the eight new towers just built in Chongqing, in China—at the urban prow where two great rivers meet—that have redefined the core of one of the world's largest cities.

The second and third floors consist mainly of open drafting areas, interspersed with models—projects built but also projects that, for some reason, never came to fruition, such as Columbus Circle, in New York City, a bewildering story; and the National Museum of China, a promising venture that fell apart for reasons I still don't quite understand. The drafting room is now an anachronistic name for a space filled with computers, where no one except old-timers like me "drafts" with a pencil or pen. Almost everyone at Safdie Architects works in this common space, subdivided by low partitions. The computers in some cubicles have

Our offices in a former factory and warehouse in Somerville, Massachusetts, just over the line from Cambridge.

My wife, Michal
Ronnen Safdie,
1973. The magazine
was marking the
State of Israel's
twenty-fifth year.

two or three screens. I happen to love physical models—to me, they are in fact essential—but digital drafting, where designs can be rotated and flipped and manipulated every which way, is how almost all the drawings in the office are created.

My own office, where I can hold meetings and small conferences—private except for the raccoon—is on the third floor. One wall is filled with books and little toys inspired by our projects, such as a Lego version of Marina Bay Sands. Behind my desk, the wall is covered with photographs of children and grandchildren. My daughter Taal and her family at Machu Picchu. My son, Oren, and his family in Egypt. My daughters Carmelle and Yasmin in Jerusalem. My wife, Michal, on the cover of *Newsweek*, back in 1973. There is a photograph from the day I showed the historical museum at Yad Vashem to Barack Obama, then a young senator. And a cartoon from *The New Yorker* poking gentle fun at Habitat '67. I spend most of the day not in my office but walking from desk to desk, meeting one team after the next, sharing sketches I may have worked on the night before or brought back from a trip and perhaps sitting with an associate by a computer screen and watching as ideas are developed in three dimensions.

The topmost floor consists of a very large conference room from which you can see the skyline of downtown Boston. The room can seat forty people. We built it when we were working on Marina Bay Sands, having learned something important about cultural variety and the different ways of doing business around the world. For giant Asian projects—like Jewel, like Chongqing—clients will fly in people from everywhere for workshops. There is a hierarchy that must be respected. There are people who must be included in any meeting. Elements of process may be as sacrosanct as they are unclear. It is not like brainstorming on the back of a napkin in a coffeeshop.

Today we are meeting with people from yet another unfamiliar culture—a team from Facebook, now called Meta, which has flown in from Menlo Park, California. We have just finished a large model for the new complex that Mark Zuckerberg intends to build at the corporate-headquarters site. The model for Project Uplift, as Meta/Facebook calls the effort, is at a scale of one inch equals twenty feet, and physically about seven feet long by four feet wide. The project was the outcome of a design competition completed only three months earlier. The challenge was to find a way to connect the two existing corporate campuses with a proposed third campus. The area in between wasn't empty—it encompassed highways, wetlands, and some urban development—and somehow had to be bridged. And the new campus had to be multipurpose, including not only an office complex but also a hotel and a town square. In the end, our team won the commission, producing a design that includes an elliptical park hovering above the terrain and connecting major structures, and also a domed garden setting, the Forum, for conferences and public programs. Now, months into the process, two presentations have already been made to Zuckerberg. Both meetings were short, preceded by intense planning and targeted on the very narrow open slots in his schedule. In a conference room, the models and drawings were laid out on tables. Zuckerberg, in his gray sweatshirt, took it all in and focused on the essentials.

Today, with the company's leadership team, we are discussing in more detail the program for the Forum, as well as issues that will need to be addressed in order to obtain the complex approvals needed from local and state agencies in California. Getting approvals—negotiating a design through the maze of building codes, zoning laws, and community boards, which differ from place to place but are byzantine everywhere—is a big and unglamorous part of what any architect does. And it is a reminder that architecture does not exist on some abstract, ethereal plane

Client meeting at our offices, 2019: reviewing the design for Meta/Facebook's new campus, to be built in Menlo Park, California.

where demigods wave their wands. It is grounded in actual places filled with actual people.

The Meta/Facebook meeting ends after lunch. I spend the afternoon in the drafting areas to meet with various teams. Some are working on construction documents for a medical school we are building in São Paulo, Brazil. We have just received samples of materials from the site, and there are details to review of the vast glass-and-steel roof structure being developed and manufactured in Germany. Another team is working on a new apartment complex in Quito, Ecuador. We have already presented and obtained approval for the overall concept and are now beginning what is known as the schematic-design phase. New input from the engineers has been received, with various alternatives for how we might refine the geometry. Across the room, our interior-design team is assembling samples and color swatches of carpets, tiles, and woods for the hospital we are building in Cartagena, Colombia. We then review the patient rooms, waiting areas, and restaurants.

By five o'clock, I am back at my desk. Emails have begun to accumulate, and some need an immediate response— queries from our field offices in Singapore, Shanghai, and Jerusalem. At any given moment, our firm has projects worth

Marina Bay Sands, immortalized in Lego.

several billion dollars in some phase of development. The projects are spread across twelve time zones. Somehow, all of this is managed with a staff of eighty in Boston and twenty to thirty, all told, at the field offices. I respond to the emails as needed, by degree of urgency, often adding sketches to elaborate a point or to explain my larger thinking. By seven o'clock, I'm at home, earlier than usual, for a quiet dinner with Michal. I report on the day's events; Michal reacts and advises. She responds with joy to the good stories and tends to take disappointing news harder than I do. Our lives and our work—Michal is a photographer—are totally intertwined.

Friday: the last day before a marathon trip. As always on the final day, there is a rush to review everything that must be done. We are working intensively on a presentation that I am to take to Singapore—to be given to the Urban Redevelopment Authority (URA) for a major addition to Marina Bay Sands. The 3D printer and the model shop are in overdrive, completing the model and preparing it for the trip. Almost everyone has seen architectural models, but few pause to consider what is involved in safely sending a big model halfway around the world. We not only need to construct crush-proof containers but also to fit them out with large windows to satisfy the security screeners and make clear to baggage handlers that the contents are indeed delicate. The art of packaging is itself a form of architecture.

For me, ever since my earliest days in architecture school and then my apprenticeship with Louis Kahn, the model shop has always been a magical place. At our firm, seven professionals toil away full-time making large- and small-scale models: the study models, the presentation models, the small mock-ups. Some building models are so big you can stick your head inside and see the interiors. Other models re-create an entire urban district or rugged rural landscape, showing a proposed building in context. Looking around the model shop, I see saws and drills, paint booths, 3D printers, laser cutters, and sheets of wood and Styrofoam and multicolor plastic. Cabinets with scores of drawers hold tiny figures of people at different scales, and tiny escalators at different scales, and tiny cars and boats and airplanes. There is material for making manicured lawns and desert scrub, fresh water and ocean, and trees and bushes of every kind: Aspen. Oak. Maple. White pine. Palm. In our projects, sustainability and the integration of architecture, site, and plant life demand that a model fully describe both the building and the landscape. The little drawers are full of material that helps us capture this ambition.

By evening (seven p.m. in Boston is eight the next morning in Singapore), we have a Webex conference call about the Marina Bay Sands addition. With this form of global videoconference, we can project images for all the participants to see, and during the meeting we can draw on the images and revise them in other ways. Later, when the pandemic lockdown took hold, in the spring of 2020, video-conferencing would keep our projects worldwide moving forward on schedule. This conference call today involves the client team, our associate architect in Singapore, and the engineering teams we are working with, based in New York. We discuss how the meeting will go on Tuesday in Singapore: what to watch out for, issues that might be raised, lingering challenges.

The next day, Michal and I fly to New York to catch a Singapore Airlines flight to Changi. Between flights, we sneak in a couple of hours with our two-year-old grandchild, Gene, Carmelle's son. The trip to Singapore, door to door, is a long one—close to twenty-four hours. I settle into the nonstop eighteen-hour flight from Newark, New Jersey, which consumes most of Sunday. As always, I do a bit of work, a bit of reading. I make sketches in my notebooks—a practice I've followed for sixty years. I have amassed more than two hundred of these notebooks, sequentially numbered, going back to the early 1960s. On long flights I generally watch a movie. There are many foreign films available on Singapore Airlines, and I always look for European or Asian films that I'd be unlikely to encounter at theaters in Boston or even New York. On this trip I watch a Polish film, Paweł Pawlikowski's astonishing *Cold War*. I sleep for at least eight hours—which is essential, because I will have to start work immediately upon landing. I have the necessary luxury of flying first class and—having spent decades traveling in seats you could not sleep in, even when traveling first class—am grateful that these long flights now have comfortable sleeper seats. Some even offer suites.

We land early Monday morning, Singapore time. I always make sure a pool is available in the hotels where I stay, because I try to swim every day. During the warmer months in Boston, and even some of the not so warm ones, such as late September, I swim in the mornings at Walden Pond, in Concord—the setting is beautiful and the discipline important. On this trip, after a swim at the hotel, I prepare for the meeting with the Urban Redevelopment Authority. Then, at noon, we hold a rehearsal. By two o'clock we have convened with government officials—presenting, discussing, getting input and suggestions. Our firm and the URA have a long history of successful experience together. We share the agency's belief in the role of urban design—to bring together multiple projects to form cohesive urban districts. It is a mutually respectful relationship, and the meeting is a good one.

In the evening, Michal and I have dinner with my client and by now close friend Liew Mun Leong, formerly chairman of CapitaLand, Singapore's largest developer, for whom we have undertaken several megaprojects. Liew at the time was the chairman of Changi Airport, under whose auspices we completed Jewel, as well as of Surbana Jurong, Singapore's largest architecture and engineering firm, for which we are designing a new headquarters building. Whenever I am in Singapore, one evening is devoted to catching up with Liew. We try different restaurants. The conversations range widely, from the state of the world to the various activities both of us are involved in. We often discuss books we have read, recent trips, and political developments in the United States, Israel, Singapore, and China. Liew is fit and energetic—he is a runner—and he speaks in rapid bursts. I must concentrate to take it all in.

On Tuesday, we are back at the airport for the six-hour flight to Chongqing, a city in the People's Republic of China that few people know much about, even though the municipality as a whole has a population of 33 million. Chongqing is also the site of the largest project that our office, so far, has designed and built: a 12-million-square-foot mixed-use complex known as Raffles City, on the historic and strategic site of Chaotianmen Square. This is where the Yangtze and the Jialing Rivers converge, forming a wedge somewhat like the tip of lower Manhattan—the Emperor's Landing, as

Raffles City, in Chongqing, China, 2020—eight towers on the "prow" known as the Emperor's Landing.

it's known. It is the most symbolically charged site in the city. The project was the outcome of a design competition we pursued jointly with CapitaLand, which is vigorously engaged with development in China.

The term "design competition" sounds straightforward, but competitions are time-consuming, expensive, and unavoidable. In a firm such as ours, more than half the projects we undertake are obtained through some type of competition, compelling us to butt heads against firms whose principals are peers and in some cases personal friends—Frank Gehry, Norman Foster, Renzo Piano, and others. Most of these are invited competitions, in which a limited number of "short-listed" firms are engaged, usually with some sort of stipend. But the stipend may cover as little as 10 percent of the cost of doing the actual work. For one major museum project in China, for which we proceeded through four competition stages before ultimately losing out, our firm spent approximately $1 million. The larger and more complex the project, the greater the risk and the expense. It is almost impossible for a medium or large firm, given its overhead and expectations, to participate in a competition for less than a few hundred thousand dollars. The uncompensated outlay can exceed several millions of dollars when dealing with, say, a competition for a major international airport. And the dynamic can be insidious: the more ambitious the firm, and the more tempting the project, the more leverage the client has in getting that short list of architects to invest more and more of their own resources in the competition. We win about half of our submissions, but that statistic is misleading—we'll sometimes have several winners in a row and then a baffling sequence of losers, causing dejected spirits and financial stress. I often wonder how the renowned architect Zaha Hadid endured years of rejection before emerging into a thriving career—a mark of her deep and sustained conviction.

Happily, we had come out on top in Chongqing. Now, after eight years of design and construction, the project is nearing completion. For several years, I had been traveling to Shanghai or Chongqing every six to eight weeks, reviewing the details, the mock-ups, the materials. In Chongqing, I have walked the site on more hot summer days than I can count, in temperatures often reaching one hundred degrees. I have walked the site on frigid winter days too. Chongqing's extreme and variable climate is the reason we decided to create an enclosed conservatory—which links four of the eight skyscrapers, on the fiftieth floor—rather than an open sky park, like the one that creates the green whoosh of a parkland plateau atop Marina Bay Sands. Think of the conservatory as a "horizontal skyscraper"—a thousand-foot-high tower lying on its side. Imagine if the Empire State Building, the Chrysler Building, the Woolworth Building, and 30 Rockefeller Center were clustered together in lower Manhattan and connected by that horizontal skyscraper—an enclosed version of the High Line in a tube, eight hundred feet above street level. The conservatory has been acknowledged by Guinness World Records as the world's longest sky bridge. Here, at the prow of Chongqing, we are creating a new kind of urban canopy, a new way to mitigate density by

The conservatory in Chongqing, a landscaped sky bridge.

means of a parklike resource up in the sky, linking people in buildings who would otherwise be many altitudes and elevator trips apart.

We land in Chongqing at one o'clock and are off to the site immediately. We have two days and a lot of territory to cover. We must review the retail podium, the apartment towers, the offices, the hotel, and the conservatory, now nearing completion. At this point, we are mostly down to fine-tuning the last details—adjusting the lighting, checking the signage, improving the interiors. All the big decisions were made long ago. The elevators have been installed, and to everyone's relief, much of the space is already air-conditioned. But it is essential to inspect in person. On-site, my eyes are wide open, like an owl's, alert for what others may not have observed and for new conditions that need watching. For instance, there is construction dust on the leaves of the trees in the conservatory—an eighth of an inch, threatening thousands of newly planted trees and shrubs. An army of cleaners must be summoned to sponge every leaf.

That night, we have a team dinner at a large round table, and the client and many of the consultants join the architects from our firm who have been on the site for the past five years. The next afternoon, on Wednesday, with our work at Chongqing finished, we fly to the gleaming city of Shenzhen, on the Chinese mainland near Hong Kong. Two weeks ago, at the opening of Jewel Changi, we had met with Shenzhen's vice mayor and the chairman of its airport authority. They had in fact brought an entire delegation with them. Word had gotten out about Jewel: that it had changed the paradigm of airport design. Americans, in particular, may not realize it, accustomed as they are to accepting the dreariness and inconvenience, and sometimes outright squalor, of many substandard airports, but in Asia, airport design is highly competitive—a fierce battle for supremacy. Singapore's Changi has enjoyed the presumptive status of "best airport in the world" for several years—the designation is bestowed by Skytrax, the international air-transport rating organization—and Jewel is a facility that no other airport can match. But Shenzhen would like to try. The city recently completed the construction of a much-photographed new airport, designed by the Italian architect Massimiliano Fuksas. The meeting in Shenzhen concerns the possibility of creating a Jewel-like complex—but even bigger—to go with it. Shenzhen was China's first "special economic zone." Its municipal administration is considered enlightened. It is also ambitious—even by Chinese standards, which have few apparent boundaries.

Upon landing at Shenzhen, we are received at the airplane's door by the airport's chief executive officer—always a good sign—and escorted immediately to a banquet. We are joined by Jaron Lubin and Charu Kokate—a partner from our firm's Singapore office—who arrived several hours earlier. Over Cantonese delicacies (duck, abalone, dumplings) the airport's CEO and its chairman, along with their team, describe what they have in mind: the scale, the program, the vision.

The next morning, Thursday, we tour the airport and the site of the proposed complex. It is to be an airport city—a node of high-end offices and high-tech research centered on our own catalyzing project, which we have come to refer to as the Crystal. Then there is another banquet, this one joined by officials from Shenzhen. It is lively and jovial, the traditional toasting fervent, frequent, and prolonged. Fortunately, the toasts do not involve the Chinese spirit Maotai, which is very strong. Perhaps in deference to a visitor, or to the hour, the toasts merely involve sips of Bordeaux. In the afternoon, we are taken around the city, a place that has evolved from Hong Kong's backyard hinterland to a metropolis of 12 million in two generations. China never ceases to elicit amazement, just as its governance and bureaucratic systems never cease to produce bafflement. But from what we have learned during our two days in Shenzhen, the airport project seems to be ours.

This is good news, but the real work and the real risk lie ahead. Design aside, the economics of architecture are daunting. In the United States, beginning decades ago, antitrust legislation forbade architects from establishing a standardized schedule of fees. Henceforward, each architect had to negotiate an overall fee—in advance—with every new commission. To appreciate the extent to which this does not make sense, consider the legal profession, where fees are generally charged by the hour at rates that cover cost and overhead (and of course profit). In contrast, the architect's fee, fixed at an early stage of the project, sets two

The raccoon in my
office window.

contradictory objectives. Carrying out the work responsibly requires that you devote to it all the hours it takes, including oversight of construction. At the same time, the business manager at the firm is always watching the clock.

The scope of a project like the Marina Bay Sands development was in the billions of dollars. Our overall design fee included fees for local architectural firms playing supporting roles, as well as for multiple engineers and specialists on matters ranging from landscape to graphics to acoustics—by themselves totaling in the low nine figures. The number of people involved from some of these architectural and engineering firms could be as many as fifty. The duration of construction was set for a certain number of years, but unforeseen circumstances could easily produce unexpected delays. Hurricane Katrina once forced our team to rethink the design of a federal courthouse we were planning in Mobile, Alabama. Inflation and fluctuations in exchange rates can have corrosive consequences. When we embarked on Lester B. Pearson International Airport's Terminal 1, in Toronto, for instance, we had a contract payable in Canadian dollars. At the start of the project, a Canadian dollar was worth ninety American cents. At the finish, it was worth sixty-five cents, cutting mightily into our fee. Virtually every component of a project like Marina Bay Sands—or Chongqing or Shenzhen or the Salt Lake City Public Library—turns out to be, in essence, a variable: the time, the people, the materials, the costs, and the state of the world, including during a pandemic. And yet, the architect is expected, in advance, to commit to a fixed fee and to bear the loss if one or more variables spiral out of control. You can set contingencies and safety factors, and sometimes renegotiate, but given the range of imponderables, architecture is a high-risk profession.

It is also a deeply satisfying one. The results make a difference in people's lives: the way people work and sleep and travel; the way they consume the planet's resources; the way they derive aspiration and inspiration from the built environment around them. There are few greater pleasures than visiting a project that is fully occupied and functioning as planned, and hearing from residents or workers about how their lives have changed for the better.

In the afternoon we make the two-hour trip from Shen-zhen to Hong Kong airport to catch Cathay Pacific's direct flight home. It is Saturday in Boston when we land. The time is eleven p.m. A week has elapsed since the trip began. I take Sunday off, catching up with friends, listening to music. On Monday I drive once again the five minutes from my home to my office.

The ivy rustles. The interior beckons with light. We pick up where we left off.

Haifa, from the slopes of Mount Carmel, in the 1940s—the city I knew as a boy. The view is similar to the view from our apartment.

CHAPTER ONE

A House on a Hill

As I reach further into my eighties, traveling around the planet several times a year, I think of the long journey that brought me improbably from the city of Haifa, before the State of Israel even existed, to the adventure in architecture I have pursued for six decades.

Haifa, now in Israel, was during my childhood administered by the British under the Mandate for Palestine, conferred by the League of Nations. Indeed, the British headquarters were in Haifa. The city lies at the southern end of a long crescent bay, with Acre anchoring the far end and the Lebanon mountains visible to the north. Haifa rose along Mount Carmel in layers. On the waterfront was the port, built and controlled by the British. It extended eastward toward the bay, where the oil refinery of the Iraq-Mediterranean pipeline, with its pair of iconic chimneys, had its terminus. Along the port was the main downtown boulevard, then known as Kingsway, now Independence Road. It accommodated the banks and business district. Immediately behind this was the lower city, or the Old Town, as it was known, its architecture mostly Arab-Mediterranean vernacular, with narrow streets and crowded markets. A wafting aroma of spices and meat roasting on wood fires filled the air. The architecture was stone—warm and Mediterranean, vaulted and domed. Since childhood I have loved domes. There is a spiritual element, I am sure—circularity symbolizes unity—but the practical, evolutionary aspect, which I would learn about only later, inspires awe in its own right: in desert regions without many trees to provide wood, domes built of brick or stone are the only means to span a large room.

Upward from the Old Town, then as now, the city changed color and character as it rose in elevation. Midway up the slopes of Mount Carmel was a neighborhood, Hadar HaCarmel, of white, Bauhaus-style buildings. At the center of Hadar HaCarmel was the campus of the Technion—today, Israel's equivalent of the Massachusetts Institute of Technology. In contrast to the stark, modern simplicity of the Bauhaus style, the Technion was a domed, symmetrical winged structure built with buff stone arcades—a European attempt at Romantic Orientalist architecture. Farther up the hill were the Baha'i Gardens, the holiest place for the Baha'i religion, where its founder is buried. I thought of these gardens in my youth as the most beautiful place in the world—the embodiment of paradise. Finally, at the crest, with extraordinary views over the harbor and the city, stood a precinct of landscaped villas and low apartment buildings. The population on the summit was dominated by highly educated and sophisticated German-Jewish immigrants—professors, professionals, businesspeople—whom the less rarified referred to as *yekkes*, a term derived from the acronym for the Hebrew phrase "fails to understand." They were made fun of—for instance, because of their accent when they spoke their newly acquired tongue—but at the same time deeply respected. They were the ones, together with other Europeans from Eastern Europe, who had brought Bauhaus architecture to Haifa and Palestine in the first place. Classical music could be heard as you passed their open windows on a warm afternoon. The crest of Mount Carmel was thick with pine trees. To this day, every time I smell pine, I think of Haifa. It remains a beautiful city.

I was born in Haifa on Bastille Day 1938. My parents had met in the city a little over a year earlier. My father, Leon Safdie, had come from his hometown of Aleppo, in Syria, in 1936. He was the ninth of ten children in a Jewish merchant family. It was never clear to me what had brought him to Haifa. Was he a Zionist? Was he looking to carve out his own turf in a family with too many competitors? But there he was, setting up his own trading business. He imported textiles, quality woolens, and cottons from England, and fabrics from Japan and India for the local markets. My mother, Rachel Safdie, née Esses, was English, but she too came from a Jewish family with roots in Aleppo. Her father's business was also in textiles, and he and his family had emigrated to England at the turn of the century, settling in Manchester, which at the time was a global hub of textile manufacturing. My mother was born in that city and brought up as a good English girl. She had a strong Mancunian accent all her life.

In 1937, my mother, age twenty-three, had embarked from Manchester on a trip to visit her sister Gladys, who lived in Jerusalem. She had sailed from England to the port of Haifa, intending to make her way overland from there to Jerusalem. Disembarking, she almost immediately met a young man who had an office near the docks. She married him a month later. With a world war breaking out not long afterward, a decade would pass before she returned to Britain. My father spoke

Arabic and a poor Hebrew and had been educated in French, and so was fluent in that language; my mother spoke English and also knew a little French. They did not at first even have a strong common language. I was the first of their children; two more, Gabriel and Sylvia, came within a few years; and a fourth, Lilian, arrived when I was eighteen, and the family was living in Canada.

The Jews of Aleppo were broadly known as Mizrahi, or Eastern Jews, but most of them also claimed to be Sephardic—that is, descended from the Jews expelled from Spain and Portugal by the Inquisition at the end of the fifteenth century. (*Sepharad* is the Hebrew word for "Spain.") In fact, the Aleppo population also included many Jews whose ancestors had never left the Middle East. The Sephardim spoke Ladino, a Spanish dialect, though in time the old Middle Easterners took on Ladino as well, making the distinction between the two groups difficult to ascertain. In the world of Jewry, strongly divided into Ashkenazi (European) and Eastern Jews, the Jews of Aleppo, no matter what their actual origin, belonged to the Eastern group. My own family came originally from the town of Safed, in Galilee—hence the surname Safdie, variants of which

The Technion, in Haifa, designed by Alexander Baerwald and completed in 1924—a flagship of Jewish education in Palestine.

(Safdié, Safadi, Safdi) can also be found among Muslim, Christian, and Druze families. Sometime during the sixteenth or seventeenth centuries, driven by the economic decline of Safed, many Jews had moved north from Galilee to Aleppo.

In Haifa, my family lived at first in a three-story Bauhaus-style apartment building in the bourgeois neighborhood of Hadar HaCarmel, home to much of the city's large Jewish community. I went to kindergarten on the campus of the Technion and recall being allowed to walk there by myself at the age of four or five, my mother watching from the balcony as I crossed Balfour Street, thus entering the campus, until I was out of sight. At this early age I already enjoyed a sense of true independence, a feeling that would become even more pronounced in my teens—an experience familiar to many Israelis of my generation.

The onset of the Second World War provides some of my earliest memories. On Friday evenings, we hosted Jewish soldiers from the Australian Army for Shabbat dinner. We went down into the air-raid shelter in the basement of the apartment building almost nightly. I remember looking with excitement at the giant silver barrage balloons that were launched above the bay, tethered to cables that served as obstacles to air attack. I remember stone towers along the waterfront emitting smoke to camouflage the oil refinery, which was the main reason Haifa was an enemy target to begin with. Luckily for us, the enemy wasn't the Luftwaffe—it was the Regia Aeronautica, and while the Italians did score a couple of hits and inflict some damage, they never landed a crippling blow.

Unlike Haifa, Jerusalem was never bombed during the war. My mother's sister Gladys lived in Jerusalem with her husband and four children. To escape the dangers of the coast, my own family stayed there for an extended period in 1940, during which my brother, Gabriel, was born. Traveling from Haifa to Jerusalem was a major undertaking in the early 1940s—four or five hours by bus on dusty roads, stopping now and again for breaks and refreshments at little shops offering falafel, orange juice, and sandwiches.

I did not enjoy staying with my aunt and uncle. Their home in Jerusalem was small and therefore crowded, and they were very religious and deeply observant, which my own family was not. They were also aggressively zealous. My uncle, an imposing figure, once offered to give me a pound sterling if I promised not to use electricity on Shabbat for an entire year—this was back in a day when a pound was worth four dollars. I can still see the silver coin in his hand. But I declined

The commercial center of Haifa, 1940. An influx of modernist architects shaped many parts of the city.

the offer. Judaism and spirituality weave through my life in important and distinctive ways, but the implacable religiosity of my aunt and uncle's family—their attempted imposition of orthodoxy—served only to fortify my resistance.

Jerusalem itself I loved, especially the Old City, with its narrow passageways and up-and-down steps everywhere. I wandered the souks, with their spice counters and local crafts. The almost orchestral hum of languages and dialects was matched by the diversity in forms of dress—Muslim women covered modestly in many colors, British soldiers in khaki shorts, religious Jews in black, English women in floral hats. The sonorous tolling of church bells punctuated the Muslim call to prayer—this at a time when the call to prayer still came from an actual muezzin in a minaret—a man of flesh and blood—rather than an amplified recording. The Western Wall, the last surviving remnant of the Second Temple, was at the time accessible only by means of an alleyway, perhaps fifteen feet wide, sandwiched between the Temple Mount and the Mughrabi (or Moroccan) Quarter. For a Jew, visiting the Western Wall in effect meant negotiating a canyon-like passage through an Arab neighborhood. The tension was palpable. One never knew when trouble would start—stone throwing, sometimes worse. But throngs of Jews could always be found praying at the wall—men mixing with women, unlike today. Because the canyon offered no perspective, hemmed in as it was by the Moroccan Quarter, the wall loomed high above one's head; there was no vast plaza, as there is now, to shrink it down to size. Outside the ancient city walls was an emerging modern downtown. The elegant King David Hotel, its tall Sudanese doormen resplendent in red turbans, overlooked the Old City. From the terrace in back you could see the city walls, the golden Dome of the Rock, and beyond the city itself the Mount of Olives. Ben Yehuda Street bustled with cafés and shops. Jerusalem to me felt very cosmopolitan. The British presence was inescapable— officials, soldiers, businesspeople. So was the old Arab aristocracy. Jewish life revolved around the Hadassah Hospital, the Hebrew University, and the Jewish Agency, the precursor of the government of Israel.

To go from the coastal plains to the higher interior elevation offered a thrilling sense of ascent to Jerusalem; the biblical language of the psalms—"going up" to the house of

the Lord—had it right. After Israel achieved independence, in 1948, and for two decades thereafter, until Israel's victory in the 1967 Six-Day War, Jerusalem was divided, and the Old City was inaccessible to Jews from Israel. But in the early 1940s, when I first knew it, the city was a unified place. As a rule, we went to Jerusalem twice a year. Every summer, we also traveled from Haifa in the other direction—north to Lebanon. Much of our extended family lived there, having moved to Beirut from Aleppo. Among other things, we would visit the resorts in the Lebanon Mountains, places like Dhour Shweir and Bhamdoun. My sister Sylvia was born in Lebanon during one of those trips—another extended stay prompted by wartime conditions. This time the worry was close to panic, though I was aware of the fact only in retrospect. The panic had to do with the advance of German general Erwin Rommel's Afrika Korps across the Libyan Desert, toward Egypt. Would Palestine be next? We stayed in Lebanon until the battle of El-Alamein, in Egypt, in 1942, when English forces under Field Marshal Bernard Law Montgomery blunted the direct military threat posed by Nazi Germany.

* * *

Our last trip to Lebanon came in the late summer of 1947. I knew that, on one level, good things were happening in the family. There was a palpable sense of affluence. The war years had turned out to be good for business, as my father was able to continue importing goods that were shipped from India by my uncles. My parents bought an apartment building on Mount Carmel. We went to Beirut and bought a car. And it wasn't just a car. It was a 1947 Studebaker Commander—the first car that extended the trunk back horizontally, like the hood. People would look at it and wonder which direction it was meant to go. The Studebaker, with its radical new design by the brilliant Raymond Loewy, was ivory in color and sensational to behold.

On this 1947 trip we ventured farther than usual, to Aleppo itself, the only visit I ever made to that city. Our extended family, children and adults alike, piled into three Chrysler limousines and drove north from Beirut to see my grandmother Symbol, fragile and petite. We visited the ancient citadel of Aleppo—an imposing medieval structure on a site that had been fortified for four millennia. Memorably, my uncles panicked when they realized that my brother and I had badges from our school sewn onto our shirts. The badges showed the profile of the Reali School, which we attended, and were embroidered with the words *vehatznea lechet*, meaning "Proceed with humility." Because the words were written in the Hebrew alphabet and marked us as Jews, the adults were fearful of letting us into the streets. Tensions ran especially high in the Arab world during those last days of the British Mandate for Palestine. The countries of the newly established United Nations were at that moment deliberating whether to form the State of Israel, which by its nature would have life-changing consequences for everyone in the region. In the end, my uncles ripped the badges off our shirts and told us not to speak Hebrew in public. We fell back on English, our other language.

Advertisement for the Raymond Loewy–designed 1947 Studebaker Commander that my parents bought in Beirut.

Palestine had been in a state of civil insurrection for several years, as Jewish fighters for independence waged a guerrilla campaign against British authorities. Our family was in Jerusalem in July 1946 when the Irgun, the militant underground Zionist organization that believed in violence as a tool of persuasion, exploded a bomb in the King David Hotel, where the British maintained their Jerusalem head-quarters. I was a child of eight, standing with my cousins at the Jaffa Gate at that very moment, looking west. We saw the flash, and a second later the horrifying sound reached our ears. Scores of people—British, Arabs, Jews—were killed.

The United Nations agreed to the creation of an inde-pendent Israeli state on November 29, 1947, partitioning Palestine, at which point the conflict with Britain broadened into a conflict between Jews and Arabs. By springtime, the battle of Haifa was underway, as Jewish forces—the Haganah, the nucleus of the Israeli military—sought to gain control of the strategically important city, which was in the area desig-nated for a Jewish state. From our home high up the slope, overlooking the port, we could hear the shooting. After stray bullets came through a window into my bedroom, we erected defensive steel plates on the veranda and windows. It is said that, after the battle of Haifa, the message from the Jewish victors to the Arab population from loudspeakers mounted on cars was encouragement to stay in the city—which em-phatically was not the message in other places during the War of Independence. And, in fact, quite a few Arabs stayed. My parents had Arab friends. Haifa today is probably one of the more successful mixed communities of Israeli Jews and Arabs. But tens of thousands of Arabs, understandably fearful, left the city in 1948, fleeing to Arab towns farther north and to Lebanon and beyond.

My recollections of this exodus are firsthand and trou-bling. With my friends, we watched as much of the Arab population picked up and left—and then watched Jews from the poorer abutting neighborhoods enter the Arab neighbor-hoods and start to loot. They would go into houses, pull out drawers, and empty the contents. People had often departed very quickly, leaving silverware and other goods behind. I saw someone take off with a stamp collection. At this young age, I already felt that life was taking a turn toward some-thing more complex.

On Independence Day—May 14, 1948—when I was not yet ten, I remember going to downtown Haifa with friends. None of our parents were with us. We joined the crowd outside city hall, listening to the broadcast as David Ben-Gurion, in Tel Aviv, proclaimed the State of Israel. Arab armies from neighboring countries would invade the next day. What is remarkable to me, looking back, is that in a period of ongoing hostilities, a bunch of ten-year-olds had the freedom of the streets without supervision.

The Reali School was an elite institution that had been founded before World War I, and I attended the outpost of the school atop Mount Carmel instead of the original school near the Technion campus. My parents had apparently had a hard time getting me into the Reali School; they once confided that I had initially been turned down, they believed, because we were Sephardic. As my father told the story, he went to the manager of the Anglo-Palestine Bank, with whom he did business, and had him intervene to get me admitted. I was one of two Sephardic students in the school; this number would swell to three and then four as my siblings Gabriel and Sylvia enrolled. As a young man I never experienced any kind of singling out, much less outright discrimination, on account of my origins. But my parents clearly had a different experience and were acutely sensitive to the issue.

At school, in my early adolescence—I was perhaps thirteen or fourteen—I did not behave as if I were there at anyone's sufferance or owed anyone a particular level of performance. I was part of a high-spirited group, verging on wild, and a frequent source of trouble. One of my report cards characterized my behavior with the words *Moshe lo sholet berucho*—"Moshe is unable to control his spirits." My mother was continually being called in to answer for my actions or at least be informed of them. On one occasion, a student was expelled for some infraction, which the rest of us perceived as a great injustice. We organized our class outside the school building, and we took stones and stood in a row and broke every window on the facade. All the parents were summoned to the school and ultimately paid for the repairs.

At around the same time I entered the Reali School I also joined the Scouts, as did virtually everyone else I knew. It was not the Boy Scouts, just the Scouts—a coed group, and one of three or four major youth movements in Israel. Others included the extreme-socialist group Hashomer Hazair and the moderate-socialist group Ha'Noar Ha'Oved. The Scouts, or Tzofim, embodied socialism's most liberal wing. Participation in the Scouts, unlike participation in my school, was something I took very seriously, drawn by the camaraderie, the idealism, and the immersion in nature, and it soon became the center of gravity of my life. My parents were supportive. Being in the Scouts meant two meetings a week, one on a weekday and one on a weekend. It meant three- and four-day hikes into the mountains and other parts of the countryside. And it meant going for the entire summer to a kibbutz, for what were called work camps.

I had spent time on kibbutzim before; they ran summer camps where younger children could swim and run around, visit the cowshed, and get a taste of kibbutz life. A work camp for teenagers was very different. We lived in wooden shacks and ate at long communal tables in the collective dining hall. We rose at five a.m. and went to work, and the work was hard—jobs like digging up potatoes, working the fishponds, clearing debris, and picking plums, peaches, and other fruit. In the evening we would wander about. Often there were bonfires with singing. Those of us in the work camps would have our own events, but sometimes we would meet up with the kibbutz kids, who all lived together, away from their parents. At a work camp one truly became a part of the kibbutz community. Back then, most of the kibbutzim were agricultural; today, some of them have big industries. One thing can be said of all the work camps I attended: the settings were beautiful. Neot Mordechai, for instance, north of the Sea of Galilee near the Lebanese border, occupied a fertile valley surrounded by mountains, some of them capped with snow.

It was a foregone conclusion that, when we graduated from school, at age eighteen, my friends and I would go together into the army—in Israel, military service was mandatory for both men and women—and would register for the Nahal brigades, which you joined as a communal group dedicated to agriculture. After the army, it was understood that the group would go on to form its own kibbutz. I had decided, at around age fifteen, that I was going to study agriculture, and I was already registered to enroll in the Kadoorie Agricultural High School, a boarding school in the shadow of Mount Tabor. The institution had been founded in 1933 as a result of a bequest by Sir Ellis Kadoorie, a philanthropist whose family came originally from Baghdad. One of its graduates was Yitzhak Rabin.

The future would turn out differently for me. But despite all that has happened, my friends from this time remain a close and cohesive unit. Sixty or seventy years later, we still gather every five years. We know that we were privileged to have shared a unique moment in history. And we recognize that this group of ours was once the center of the world.

* * *

Today, as an architect, I do indeed "work the land," though not in the way I had anticipated. In those early years, I don't remember thinking consciously in terms of architecture as a subject of specific interest—and yet, looking back, I can see a connection to themes that would become central to my becoming an architect. The Baha'i Gardens, which almost functioned as my backyard, instilled a deep and enduring love of gardens and landscape. Intuitively, I was aware of the two architectural languages expressed in the middle of Haifa—the Mediterranean vernacular (stone, sensual, warm, domed, rustic) and the modernist International Style (white, minimalist, cool, curvy, formal). I certainly didn't use words like "language" or "vernacular" in an architectural sense, but I registered the aesthetic distinction between downtown Haifa and uptown Haifa. I may have taken it for granted at the time, but I was aware of the difference and it probably planted something that ripened over time.

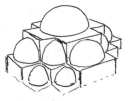

"Since childhood I have loved domes": My early sketch for Yeshiva Porat Yosef, Jerusalem, 1972.

In the early twentieth century, when what was called the Yishuv, the Jewish community in Palestine, started building places to live—new towns like Tel Aviv and new neighborhoods like the upper city of Haifa—it initially adopted a kind of romantic Middle Eastern approach, making buildings with arches and domes. The original Technion, built in 1912, is such a place. But then, in the 1930s, a wave of immigrant architects arrived from Europe, fleeing the growing anti-Semitism and the seemingly inevitable drift toward war. They were Bauhaus-trained or German-trained or Vienna-trained, and they built whole communities employing a modernist vocabulary that had not taken root to such an extent anywhere else

In Israel with friends from the Scouts, 1952, at Kibutz Hulda. I am at lower right.

in the world. Unlike in Athens or Berlin or Milan, where one finds Bauhaus buildings mixed in among older structures, in Israel, the properties being newly developed didn't have any older structures to compete with. The so-called White City, a Bauhaus-inspired neighborhood in Tel Aviv, is today a UNESCO World Heritage site. It has some four thousand Bauhaus-style buildings, mostly three- or four-story apartment buildings but also schools, concert halls, theaters, department stores. This style can also be found in kibbutzim. There were some remarkable local adaptations. In Jerusalem, the British had legislated in 1918 that everything in the city must be built with Jerusalem limestone, a softly golden local material that has been used for building since ancient times; British authorities hoped, by mandating its use, to preserve the city's harmonious color and texture. And so, in Jerusalem, Bauhaus architecture is clad in Jerusalem stone. I call it Golden Bauhaus.

I can see other aspects of my early interests that are almost premonitions of what would come. Thinking now of the high regard in which I held my father's Studebaker, I realize that I must even then have had an intuitive feel for design, although the name Raymond Loewy—responsible for the streamlined S1 locomotive and, later, the Shell Oil Company logo and the classic color scheme and typography of Air Force One—meant nothing to me.

The demands of ordinary life in Israel also brought a familiarity with what communities must have in order to sustain themselves. During and after the War of Independence and the first years of Israel's statehood, amid stringent austerity, everything was rationed: two eggs per person a week; very little meat. My family had an easier time than some, because my mother's brothers, who had moved to Dublin, Ireland, from Manchester, were shipping us boxes of food. But still, with the austerity, we were all encouraged to become farmers. And I took to that. Because not everything was built up and urbanized, we could easily use the ancient terraces around the house to plant vegetables. I had a henhouse in the garden and twenty-five hens, producing quite a few eggs a day. I had a donkey. I had pigeons.

I also kept bees—mesmerizing for both their social and architectural dynamics. It started as a school project—the school helped order beehives, and I got an Italian breed of bees along with instructions. I was completely absorbed by the social organization of bees and also by how they come to build the structures they do with such precision. In modern beekeeping, the bees receive a little imprint of wax, but it's just the outline of the hexagonal cells—the bees themselves build the walls of the cells with wax of their own making. They would not accept wax cells made by machine—the architecture would not be precise enough, from their perspective. At the time, I wasn't thinking in terms of architecture or about the process of bringing physical structures into existence. I would later learn about geometry and close packing and the platonic solids and other relevant ideas. In the moment, I just enjoyed beekeeping.

Another school project may have had a lasting impact— an exercise that had to do with the idea of harnessing the power of nature. All of us needed to come up with a project. I and a friend—Michael Seelig, who also eventually became an architect—made an enormous model, which we built using an old door as the foundation. It was about three feet wide and about eight feet long. The surface was fashioned out of clay and plaster. We created mountains with waterfalls and painted the scenery to be realistic. Then we added hydro-electric installations and windmills to generate power. It weighed a ton. Our parents had to get a truck to take it to school. But the model was a sensation. To this day I enjoy immersing myself in the work of the model shop at the office. Fortunately, we use lighter materials than clay and plaster.

Finally, as I think about influences that may have planted seeds—flowering only later into a passion and a career—I cannot forget my own home. I was born in a three-story modernistic apartment building. Our residence occupied the second floor, and it had a small balcony. A communal staircase was shared by several families. When I was ten, we moved up the mountain, toward the crest. We lived on the third, and top, floor of a hillside apartment building, and one entered the apartment directly by means of a bridge from the garden—indeed, because the building was on a hill, every floor could have its own private entryway. We had a magnificent view of the city, and the entire rooftop was ours to enjoy.

My bar mitzvah took place on that roof in July of 1951. This was the time of austerity—*tzena*, as it's called in Hebrew—but the austerity was relieved in our household by the arrival of fruit, canned meats, and other delicacies sent from uncles in Ireland. The idea that Ireland in 1951 loomed in my mind as the Land of Plenty gives you some idea of what ordinary life was like in Israel at that time. This hillside apartment expanded my concept of "home," an ideal that was taken up a few more notches whenever we visited an aunt, Renée Sitton, and her husband, Joseph, the deputy head of the Anglo-Palestine Bank. They lived in a villa above us on the summit of Mount Carmel. The home had a garage on the street and a walled garden with a path winding through it. Gently rising with the slope, the house was one story high, with every room opening onto the garden and

In my thirteenth year, 1951. Our home inspired my later thinking about what any house should—and could—be.

views of the sea far below. It was beautifully landscaped. I
soon came to think of my aunt's residence as representing
a version of the ideal home.

In time, when I was in my twenties, I developed ideas
about the characteristics of an ideal home, and those ideas
are rooted in my boyhood. The ideal home, I concluded,
must have its own territory—well defined and private, even
if it is small. It must always have a garden or a courtyard,
or some other form of outdoor space—a transition zone,
making a connection between the sheltered world indoors
and the natural world outdoors. If possible, the home would
have a view of some kind. It doesn't have to be the Amalfi
Coast or the Grand Tetons—it can be a bit of countryside,
or a few trees, or the rush of water, or even nearby buildings
if they are the right kind of structures. Many years later, the
concept of Habitat would reflect all of this.

* * *

In the new State of Israel, my family faced some serious
challenges. My father in particular felt that his life, at
best, was being made unnecessarily difficult, and at worst, he
was coming under deliberate attack. Israel's new Mapai Party
government, socialist and dominated by trade unions, was
choking his business—denying him import licenses, showing favor to its own allies, and demanding an equity stake
in his enterprises. He had a big inventory of goods—mainly
textiles imported before 1948—but new forms of price con-
trols made that inventory unprofitable to sell. Some of his
competitors were selling their stock on the black market,
and making good money, but he would have nothing to do
with illegal transactions. He believed he was being treated
as a black-marketeer anyway—watched closely and incon-
venienced in every possible manner. He took to calling the
government "the Bolsheviks." But there was more to it than
that: my father believed he and other friends were being sin-
gled out for harassment because they were Sephardic, and also
because our relative affluence had become a source of envy.

He was also worried that I was being indoctrinated into
socialism at the Reali School and in the Scouts. I certainly
identified myself as a socialist, and I was committed to
the kibbutz movement and to the cooperative movement,

reflecting the values of Zionist socialism that were woven through our education. I marveled in those early days of the nascent state that the bus company could be owned and run as a cooperative and that various industries—glass, steel, dairy— were also cooperatives, all owned by the workers themselves. I saw a kibbutz in my future. Once, on parents' day, my father made a speech at the school about socialist ideology and why it was inappropriate for an educational institution to press such an agenda. Even before it happened, I am ashamed to say, the event filled me with anticipated embarrassment—not only for the substance of what he would say, but also because of his less-than-perfect Hebrew and his Sephardic accent.

Israel's elections did not promise relief for an entrepreneur like my father. In the first Israeli elections, in 1949, my father actively supported a small Sephardic party, which won four seats (out of 120) in the Knesset, the Israeli parliament. One of its members became the minister of police—"the token Sephardic minister," as my father put it. In the second elections, in 1951, he supported the General Zionists, a private-enterprise party, hoping it would do better and bring relief for the business community. It did do better, winning twenty seats, but not nearly well enough to unseat "the Bolsheviks of Mapai."

Increasingly desperate, my father went to Italy—to Milan—in an attempt to organize a new business. He generated plans to bring a nail and screw factory to Israel, and sent them back to Israel for government approval, I remember, rather than returning home to present them himself. In the early years after independence, emigration from Israel was discouraged not only by government rhetoric but also by practical measures—for instance, it could be very hard to get an exit visa. His suspicions may have been justified: when the trade unions saw what he had in mind, they demanded 51 percent ownership of the new business, which to him was unacceptable. It was time for my father and mother to make some hard decisions.

My mother joined my father in Italy for three months; as a British subject, she had no visa concerns. Because I was somewhat uncontrollable—rough on my siblings and generally disobedient when confronted by adult authority—I was put up with the family of two teachers from my school. An older cousin came to stay with my more manageable sister and brother. During those three months, while living with the teachers, I settled down, doing my homework and suddenly getting good grades. When a three-day hike had been scheduled and then got canceled, and some of my classmates decided to get on a train and disappear into Tel Aviv for three days, instead, for once reflecting some strange manifestation of maturity and concern for the family I lived with, I decided not to join my friends on this adventure.

In early 1953, my parents made their decision: we would be leaving Israel and putting down roots elsewhere. My mother broke the news to us as soon as she returned from Milan. Emigrating from Israel in 1953 was a big deal. Jews who left were referred to as *yordim*—"dissenters." It was a derogatory term, painful to hear. Emigration was seen by the authorities as a sort of national affront. Tickets for passage by ship or airplane had to be purchased with money sent from abroad. My teachers at the Reali School, learning of our departure, invited my mother and me to school. They acknowledged that I was the class troublemaker but said they would welcome me back if I ever returned. In the end, it seemed, Israel wanted to keep even its misfits.

The news that my family would be leaving Israel was, for me, deeply traumatic. I was firmly planted and happy. I was also living in a spacious home—not everyone in Israel was—and enjoying a comfortable life. Regardless, as a boy of fifteen, I didn't have much choice in the matter. Looking back, I can appreciate a reality that I never gave much thought to at the time: how searing the decision must have been for my parents. My father was forty-five and my mother was thirty-eight—they were young people but not "young" when it comes to changing countries. My mother began selling off our possessions. She kept the Bechstein grand piano and some of the furniture, which was put into storage until we knew where we would be living, but she sold off the Studebaker. The family business was shut down.

Financially, our situation turned out to be more precarious than we might have expected. My father had spent most of his savings as his business deteriorated, but he counted on one reserve. During World War II, as my uncles shipped merchandise to Palestine, some went to Beirut, where my father had established a partnership with a cousin. He assumed that a considerable amount of income had accumulated in Beirut and counted on the money as funds drew down in Israel. He had had little contact with the Beirut office for several years because the Israel-Lebanon border had been closed since independence. He had considered sending my mother, a British subject, to Beirut to arrange his affairs but abandoned the idea—such a trip was too dangerous and risky at that time, given her Israeli residence. Once living outside of Israel, however, he had access to Beirut once again—and discovered that the cousin had used up the money. My father's reserve was gone.

Beyond money, there was also the vexing question of where to go. Much of my father's extended family had left inhospitable situations in Syria and Lebanon and was living temporarily in Milan. Relatives had scouted out various options and had their eye on Brazil. They had gone so far as to send one family member to South America, who returned with a license to build a pulp-and-paper factory. So, Brazil was a possible destination, and many family members pooled their resources and made their way to the Southern Hemisphere. They proliferated and did well, in industry and banking. Today, there is a formidable Safdie clan in São Paulo.

My mother did not warm to this plan or anything like it. She wanted to go only to an English-speaking democracy. These words of the girl from Manchester remain vivid in my head: "an English-speaking democracy." And so, our destination became Canada. My father knew a few people in Montreal, including an older brother, who would assist with the visa. His youngest brother, Zaki, would lend him $20,000 to start a new business. Getting entry visas was going to take a while, and we'd also have to get medical tests and clearances, so we had to stay somewhere in a holding pattern until the paperwork was ready. My parents naturally chose to join our relatives in Milan.

In February 1953, we boarded a flight from Lod Airport—today Ben Gurion International Airport, much of which I have, coincidentally, ended up designing—for Rome. It was my first airplane trip. The flight attendants were dressed with what to an Israeli boy seemed like unattainable elegance; we ourselves wore khaki shirts and khaki pants on just about all occasions. (Gray pants would have been considered bourgeois.) I was astonished that food was served. Two years earlier, my father had been on an El Al flight from Lod to Paris that

At Lod Airport, 1953. Moments later we would leave Israel, bound ultimately for Canada.

had crashed on takeoff. The airplane had departed in the midst of a snowstorm, and the wings had not been properly de-iced. Miraculously, everyone survived. That episode was certainly on my mind as we took off. But once we were actually up in the air, I adjusted to the experience quickly—as if it were something I'd been doing for years.

We landed in Rome and were greeted by my father. It was strange to see him after the passage of a year—and I'm sure it was just as strange for him, if not stranger, to see us. Children grow and change quickly. At the airport, I remember going to get a snack and having my first introduction to a drink called Coca-Cola. I can still taste it on my lips. That night I ate pasta and a steak cooked rare. In subsequent days we visited Hadrian's Villa, the Colosseum, the Pantheon, the Baths of Caracalla. I loved the pines of Rome—the umbrella pines that seemed to define the horizon everywhere one looked—and the Italian landscape more generally. After traveling north to Milan, we stayed there for a month. We went on excursions to Lake Como and to Stresa, on Lake Maggiore, where island palaces rise from the water. One evening, my brother and I climbed up to the roof of Milan's Duomo. I was captivated by the cathedral's ornate Gothic wizardry—by its sheer *architecture*, a concept I may at last have been starting to reckon with. Now I understand my parents' overall strategy: to soften the blow of leaving our home in Israel, we children were being treated to three months of wonder in Europe. It was a strategy that would have a lifelong impact.

During two months of travel, we also visited Paris and London. It was intense. The architecture, museums, landscapes, and cuisine were all new to a fifteen-year-old who had never been outside a small patch of real estate in the eastern Mediterranean: the Louvre, Versailles, Notre-Dame, the great parks of London, Saint Paul's, Westminster Abbey. Madame Tussaud's wax museum was of course mandatory—I was by no means drinking exclusively from the trough of high culture. I took it all in, enjoying myself, realizing only later how transformative the experience had been.

Growing up in Israel in those early years of nationhood had been an extraordinary experience. But I recognize the insularity of the scene. Until my mid-teens, I hadn't met many foreigners, hadn't eaten unfamiliar foods. Our diet had consisted of traditional Aleppo cuisine—stuffed eggplant, roasted chicken, and of course hummus and olives. Lobsters? Oysters? These were things I could not imagine eating. I knew basically nothing about the world, except the small world I mistook for the universe. This was before television and even before glossy magazines. Then, all of a sudden, I was thrust into a new environment; indeed, into many new environments. Every day brought a fresh discovery. By the time I got to Montreal, after three months in Europe, I was a different person.

We arrived in late March 1953, greeted by the lingering gray of winter, the early darkness. The buildings were a somber brick—not the white plaster and glowing limestone I was used to. There was snow on the ground, but the seasons were changing just enough to add an element of slow

A memory of my first visit to the rooftop of Milan's cathedral remains indelible.

melting to the equation, and everything left by dogs in the previous three months had combined with the mud and the trash left by human beings to turn the streets and sidewalks into a slippery soup. We checked into the Queens Hotel, on Peel Street, for several days until we could move to an apartment on Sherbrooke Street in Westmount, an anglophone neighborhood in the bilingual city. In a few months, yes, the trees would be green. There would be warmth in the air and color in the sky. But for the moment, I was in a state of total shock.

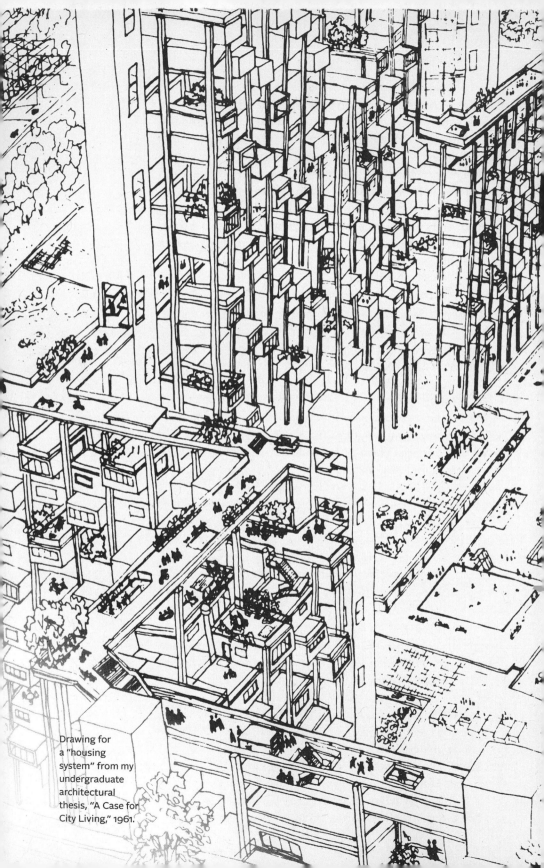

Drawing for a "housing system" from my undergraduate architectural thesis, "A Case for City Living," 1961.

CHAPTER TWO

Ideas and Mentors

Montreal is where I decided to become an architect. The city would shape me in other important ways, opening new sets of doors even as others were forever shut. If my family had not pulled itself up by the roots, I might well have followed a different path in life. But Montreal became my home, and the people I met there proved formative.

Soon after we arrived in Montreal, I enrolled in ninth grade at Westmount High School. The administrators decided that the name "Moshe" was too difficult and unusual and that "Morris" would be better. They wrote "Morris Safdie" into their records. It was already April, so there was not much of the school year left to go. My English, we all realized, was shaky. I could speak the language—after all, my mother was British—but my written English left much to be desired. The consequences of neglecting my homework in Israel were coming home to roost. To remedy the situation, my parents sent me to summer school in the Laurentian Mountains, north of Montreal. The school consisted of three or four shacks along a river. A dozen or so other students attended with me.

We were in the middle of nowhere, but the school had positive features. One of them was a girl my age named Nicole, who brought a frisson to my young life. The school also had a gramophone and a good collection of classical music. I had heard remarkably little music up to that point. We didn't have a gramophone at home in Israel, and I don't remember hearing much music on the radio. Here, in the Canadian countryside, among the trees by the river, I discovered classical music. I began with Rimsky-Korsakov, Tchaikovsky, and the operas *Madama Butterfly* and *La Bohème*. At the school for just two months, I became hooked on music.

When I returned to Westmount High in the fall, the picture brightened. My written English had greatly improved. I could write from left to right across the page almost as easily as I could write in Hebrew from right to left. My English teacher, Mr. Bernard, introduced me to *Julius Caesar* and to the Book of Job—it was my first Shakespeare and my first exposure to the King James Bible. I had studied the Bible in Israel, of course, and knew passages in Hebrew by heart. But I had never warmed to scripture and hated the heavy-handed manner often used to teach it. Now, reading the King James

Version with Mr. Bernard, I discovered the sheer beauty of Job in all its moral and philosophical complexity. I became interested in the Bible more generally and also in the evolution of Judaism. I grew interested in language itself and came to love English. And with this newfound interest, I started getting good grades, baffling everyone. Westmount High was a special place. Leonard Cohen graduated from there four years ahead of me. Future U.S. vice president Kamala Harris—who lived in Montreal when her mother accepted a research and teaching job at McGill University—graduated from Westmount twenty-six years after me.

Westmount High is also where I first encountered Jews who were not Israeli Jews, which was perplexing. The student body at Westmount was perhaps 25 percent Jewish, but these Canadian Jews were not like any Jews I had known. I was a proud Israeli who took for granted both my national identity and my Jewishness. Jewish identity was a fact, not a question, not something one had to work at. Hebrew was a mother tongue—something spoken as a matter of course. The other Jewish students at Westmount shared the outlooks and concerns found among Jews throughout North America—about being a minority, about anti-Semitism, about having holidays to celebrate or observe that were out of sync with everyone else's. Their Jewish identity was complex and varied widely. Some might have grown up in observant, traditional homes. A smaller minority, who had attended Jewish schools full-time, might speak Hebrew. Many had little deep knowledge or experience of Jewish heritage at all. They may have felt a strong bond with the Jewish community, but they were continually conscious of their minority status. There's a word in Hebrew, *galut*, meaning "exile," and my Jewish classmates, being part of the diaspora, brought the word to mind. I could not really identify with them. I became friends with some but did not socialize with them as a group.

I started looking for Israelis of my own age who were in Montreal at the time. I found half a dozen, all at different schools. I changed my name back to Moshe. I grew a mustache, which in the context of a Canadian high school in the 1950s was very odd but would not have been unusual in Israel. Meanwhile, I was corresponding furiously with my friends in Haifa, vowing that as soon as I finished high school, I would come back and join the army with them.

This plan became a source of terrible fights with my parents, particularly my father. He was working tirelessly to start a new life, and I was jabbering about going back to our old one.

One of the people in our group was a young woman named Nina Nusynowicz. She was an Israeli but had not been born in Israel. A Holocaust survivor from Poland, she had come to Israel in 1946, at age nine. Nina had not been in a Nazi concentration camp. She had spent the war years in Poland hiding in forests and on farms, and her experiences had been horrific. Her parents had been physically separated during the war, and they divorced after reuniting in Israel, where her mother still lived. Her father had moved to Canada, remarried, had a child, and eventually invited Nina to join him. She was seventeen, a year older than I was, and just finishing high school. When I met Nina, she was living, unhappily, with an aunt and uncle—joining her father and his new family had proved too difficult and contentious.

Nina was blue-eyed and blond—characteristics that might not have seemed particularly Jewish in wartime Poland and may have contributed to her survival. She had a winning smile. We became very close. My parents were not happy about the relationship. Nina was not Sephardic—"our people"—and she came from a broken home. In their imaginations, my parents wished for something different for me.

Meanwhile, in high school, to help the family, I held full-time jobs every summer and part-time jobs during the school year—first at a camera store, then at a hardware store. My father was building up a business, and our family did not have resources to spare. My father would go to the office every morning; my mother, with her typing skills and her English, would join him to handle the secretarial work. (When I was at university, she helped type my papers too.) I remember what a very big deal it was when my parents bought bicycles for me and my siblings.

As I neared the end of high school, I naturally began to think about what I would do in life. Everyone in my class was given an aptitude test, the results of which supposedly indicated what fields you might be suited for. My results suggested that because I was good at art and very good at math, architecture might be a path to consider. The idea of architecture sounded intriguing. I had never met an architect or visited an architect's office, but in my favor, I had a long history of doodling (mostly car designs). And I had never forgotten the excitement I had felt when touring the monuments of Europe. Something must have been brewing inside me. A path I had never considered suddenly seemed like the only route to take.

I announced to my parents that I would be going to architecture school. My father insisted that I join him in the family business, an age-old Sephardic tradition for the first son. I refused. To break the impasse, my father dispatched an emissary to me—someone my father worked with who could speak unemotionally and make the case for going into the business. I listened and then told the emissary that I was holding firm. So, we negotiated a deal: I would work in the business in the summers, to the extent that I could, in order to learn enough to be able to make an informed decision down the road. Meanwhile, I would attend architecture school.

The only place I applied was McGill. This was not an era of college tours. In Montreal, one didn't think about Harvard or Yale. The English-speaking university in town was McGill, if one's grades were good enough, and my relationship with schoolwork had been transformed for the better. McGill is where I went.

* * *

Generally speaking, architecture education used to be an undergraduate program that led to a professional degree—a Bachelor of Architecture, or B. Arch. In some schools, the program lasted five years. McGill's was a six-year program, and after graduation one worked for three years in a professional office and became a licensed architect. If interested in further studies, one could pursue a master's degree, but that was, at the time, above and beyond what was needed to practice. Over the years, in the United States and now also in Canada, the system has changed. Now, often, one gets a traditional bachelor's degree—in science, psychology, economics, history, philosophy, whatever—and then applies to graduate school for a Master of Architecture degree.

At McGill, when I arrived as a seventeen-year-old in 1955, the first year in the architecture program consisted largely of general education—English, mathematics, physics, and other courses of one's choice. Architecture at McGill was part of the school of engineering, and deeply influenced by that central fact: architecture was about *building*, and building required technical expertise in many fields. The atmosphere at McGill was very different from that at, say, the University of Pennsylvania and other schools where architecture came under the faculty of fine arts.

McGill's architecture school had already been transformed from one steeped in the tradition of the Parisian Ecole des Beaux-Arts to one oriented toward modern architecture and modern teaching methods. This had occurred when John Bland took over the directorship, just before the Second World War. Many components of instruction had a distinct Bauhaus influence. This postwar period is also when women were finally allowed to enroll and to teach in the program.

From the perspective of many decades, I can see that I came into architecture at a good moment. When I entered the field, in the late 1950s, the profession had been grounded intellectually and philosophically, for two generations, in a pronounced sense of social responsibility. Until the emergence of the modern movement, architecture had been an elitist undertaking. Architects worked exclusively for clients who had money, building spacious residences, palaces, churches, museums. The modern movement, pioneered by visionaries such as Frank Lloyd Wright and Le Corbusier, and by the Bauhaus school and the Amsterdam school, brought a radically new perspective: It insisted that architecture had to benefit everyone. It needed to provide housing for masses of people, not just the affluent, and to think of cities as a holistic environment, not just a locus for a few grand public buildings. It must concern itself with infrastructure—things like transportation and utilities and other services—that encompass and improve society as a whole. All of this resonated with the values I had absorbed as a youth in Israel.

John Bland was an elegant man who gave us elegant buildings. He designed the Ottawa City Hall while I was a student—a modernist building notably showing the influence of Ludwig Mies van der Rohe. His partner in professional practice was Charles Trudeau, brother of Pierre, later Canada's prime minister. French Canadians were not yet the dominant force in Montreal, though numerically they made up a majority of the city and of the province of Quebec. The upper crust, in society and business as well as in architecture, was still very WASPy. The business community was dominated by banks and corporations, such as Alcan and Air Canada, that were then rooted in English-speaking Canada.

Every architect of my generation has had what I think of as an "Ayn Rand moment"—the moment when one thinks of the architect as a self-assertive individual and an all-powerful creator with a singular grasp on Truth. I read Rand's 1943 novel *The Fountainhead* very early in my time at McGill and recall absorbing it with great excitement. The architect in the book, Howard Roark, Rand's Nietzschean hero, was, briefly, my idol. Gary Cooper's portrayal of him in the movie only reinforced my enthusiasm, though I couldn't quite see myself in Gary Cooper's image. The enthusiasm was short-lived, at least in my case. I came to appreciate that Roark was a self-obsessed narcissist, motivated by nothing noble. My ideal of an architect became Roark's antithesis. Architecture is a mission, and an architect has a responsibility (to clients, to society) that transcends the self.

It was widely believed that the model for Roark was Frank Lloyd Wright, and Wright's words and behavior certainly gave evidence of some Roarkian characteristics. When asked, in a courtroom in the 1920s, to state his profession, Wright called himself the world's greatest architect. Upbraided later for immodesty, he noted that he had been under oath. That said, I came to see that Wright was a great architect whose work possessed a true humility. He cared for the lives of those he built for.

In the summer after our first year—I was still working for my father—my classmates and I went to sketching school and spent three or four weeks drawing in towns and villages outside Montreal. Because architecture was part of the school of engineering, we also went to surveying school, learning how to measure and map the terrain with instruments. I even got a summer job at one point as a surveyor. Learning how to survey underscored the importance of the shape of the land—the subtle grades, the drainage, the overall character of the site. I also found that, with a pen or pencil in my hand, I could think visually and spatially in three dimensions, and translate that into a sketch. In time—it would take decades—I acquired the ability to think in three dimensions even without a pen or pencil and without sketching. In the series *The Queen's Gambit*, the young chess player, lying in bed, can visualize the

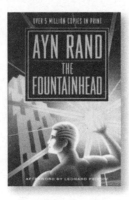

How not to be an architect: the cover of Ayn Rand's novel *The Fountainhead*, adapted into a 1949 movie starring Gary Cooper.

moving pieces of a chess game in three dimensions, as if the board is upside down above her. That image brought me a shock of recognition.

By the second year at McGill, we were getting bigger doses of architecture. And I, personally, was captivated. We had our first design studio, working with a faculty member, virtually as an apprentice, to develop actual projects. We began, each of us, by designing a small house. I won't pretend that this first effort of mine was anything special. The house sat on stilts on a slope. It had a tilted roof and big windows that opened up to the south. It was very basic. In a parallel course, we created the components of that house using real materials. This building lab was a key ingredient. It meant that, even as we were conceiving a building, we were thinking about how to actually construct it.

In architecture education today—and, alas, sometimes in practice—there's a tendency to draw something as if it won't actually need to be built, as if one is free of gravity and has magic materials at hand. One of my teachers at McGill, a legendary professor named Stuart Wilson, fought that tendency. He would say, "We're going to make a model. We're going to use pieces of wood. We're going to draw how we intend those actual pieces of wood to connect." We physically built a corner of the structure at close to actual size with wood and nails and bolts. The idea that architecture needs to be *built*, not just *drawn*, became a basic part of my understanding. Architecture is a building art. It is a language made up of physical materials. It is not about abstract concepts that are verbal, or about visual images that can't be translated into reality.

And building is not easy—not even the most basic of elements. A staircase is familiar to everyone, but making one to suit a specific purpose can be surprisingly complex. Change the height of the riser or the width of the step, or the ratio between the two, and the ergonomic act of ascending and descending is altered. We've all had the experience of being on a staircase, indoor or outside, where the width is too deep for one step and not deep enough for two.

In that second year at McGill I took a course in soil mechanics and learned how to design a foundation. A Chinese Canadian professor named Raymond Yong, who was passionate about the characteristics of the land under our feet, would give us the problem of building a ten-story structure on a certain soil type, knowing the coefficient of elasticity. He would say: "Here is a place where you must build on silt, and below it there is a layer of clay. Do you go through it or don't you?" At the time, a building in Montreal had started to tilt because of faulty foundation design and ultimately had to be taken down. It was a warning. Yong's course was as foundational—literally—as the subject matter. Buildings sit on land, and land behaves in a variety of ways. Clay is different from sand. Gravel is different from bedrock. I learned to think of architecture and structure as a single thing.

In the third year we began to have more complex design studios. Stuart Wilson ran one of them. Wilson, whom I came to know well, was a passionate designer. He was also an example of an architect who was extremely gifted but did not have the commercial instinct to run a business. Wilson's special talents lay in teaching, to which he brought commitment and energy. In particular, he drove home the sheer complexity of any building project—the countless variables that intersect from every direction. Indeed, every aspect of knowledge and life—the sciences, the social sciences, the humanities—can come into play in the course of creating a building.

While the architectural curriculum must focus heavily on knowledge-based classes—how to make a building perform; how to make sure it is safe—the studio, conducted as cases studied with faculty oversight, provides the intellectual synthesis, where one grapples with concepts previously encountered mainly in the abstract. Consider, for instance, the question of *proportion*, and the related matter of *composition*. Proportion governs virtually every step in the design process: The proportions of a facade. The proportions of a room. The relationship of width to length to height. A room that seems too small because the ceiling is too high would seem just right if the ceiling were lower. Historically, proportion has been a hotly contested and deeply investigated subject. Greek, Roman, and Renaissance architects were obsessed with the golden ratio. (In a rectangle defined by the golden ratio—like the front of the Parthenon—you can remove the portion of the rectangle equal to a square, and the rectangle left behind will also be defined by the golden ratio.) Le Corbusier had his "Modulor" ideal, based on the height of a person with an arm upraised. Manipulating proportions can make a big structure appear smaller—bunching several floors to appear from the outside as one. Or it can make a building seem without scale at all, as if woven into a seamless sky. The proportion and shape of Michelangelo's square atop Rome's Campidoglio—which is actually not a square but a trapezoid, and whose surface is in fact not level but rises gradually and imperceptibly toward the center—have everything to do with the fact that the plaza seems at once grand and intimate.

Designing in the studio exposes a student to all aspects of architecture, from the objective and practical to the intuitive and subjective.

One of my third-year classes at McGill was conducted by a Bauhaus-trained artist named Gordon Webber and was modeled after a famous course given at the Bauhaus by László Moholy-Nagy and later by his wife, Sibyl. Webber had been seriously injured in a car accident as a child, and his physical development had been impeded as a result. He was exquisitely alert to the ways we apprehend beauty and a sense of well-being through our various senses, and how our senses work in combination with one another. Webber taught a design course, and he had us construct a series of models of various shapes, lined by different materials and textures, from mirrors to swatches of fur. The idea was to explore the sensual nature of design.

Peter Collins, a British scholar, taught the history of architecture. Collins took us through antiquity, Greece and Rome, Romanesque, Gothic, the Renaissance, Bramante and Alberti, Palladio, and all the way up to the present. He was an expert on Auguste Perret, an early French modernist before Le Corbusier, whose biography he had written. Perret had been a pioneer in the use of concrete and prefabrication—I would have reason to think of him again in a few years. I was particularly taken by Perret's observation that one sign of a good building is if it looks good as a ruin. Collins taught us that architecture is a rational process. While intuition and passion play a big role, there is also a logical sequence: structure, and the expression of structure, and the materiality of structure. Collins cited and amplified the classical teachings, such as those of Vitruvius, the Roman architect-historian, whose dictum was that architecture is about commodity, firmness, and delight. "Commodity" stands for the usefulness of a building, the life intended in it. "Firmness" stands for

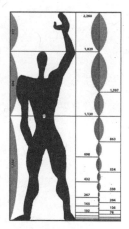

Le Corbusier's famous "Modulor," an attempt to relate human scale and architectural proportions.

structural stability, and "delight" for the beauty and pleasure brought by architecture. Above all, Collins believed, architecture needs to be thought through. It's not just a matter of "feelings." Feeling is not a substitute for thinking.

Collins not only made me think but expanded my ability to research a subject. I remember doing a paper for him on the influence of magazines such as *Progressive Architecture* on the profession—and looking back at what they were publishing in the 1930s and '40s. This was the period when Frank Lloyd Wright was creating his iconic Fallingwater, in Pennsylvania, and yet the magazines were full of classical buildings and did not give Fallingwater even a mention. I saw the influence of advertisers and the heavy hand of the fashion of the day.

I wrote another paper on public opposition to major urban projects. It was focused on the planning for Central Park, in Manhattan, and Mount Royal Park, in Montreal, which had both been designed by Frederick Law Olmsted. The paper was built on documents from the archives of New York City and Montreal. Both projects were fiercely fought over. Landowners had an obvious stake. Some questioned the very utility of vast urban parks so far from what was then the heart of the city. At a time when urbanized Manhattan stopped at around Fortieth Street, some critics argued that Central Park was a remote irrelevance. Indeed, long-term planning always faces opposition; ingenuity and commitment are needed to overcome it. This was a lesson I would find myself learning and relearning throughout my career. The forces that swirl around a project need to be understood. A purist who simply ignores them will not realize any big ambition.

* * *

By the time my classmates and I came to the fifth year, we had access to a wider range of faculty. Douglas Shadbolt, a Canadian from Vancouver, was an inspiring teacher. Another visiting presence in the school was Arthur Erickson, a graduate of McGill and the leading Canadian architect of the day. He designed Simon Fraser University in British Columbia and the Canadian embassy in Washington, D.C. But my education was not confined to the classroom. In 1959, at the end of that fifth year, I saw an advertisement for a traveling scholarship offered by the Central Mortgage and

The iconic
suburban sprawl
of Levittown,
New York—
encountered
during the
traveling
fellowship—as it
appeared in 1949.

Housing Corporation of Canada. One student from every
Canadian architecture school would be selected to travel
across the United States and Canada for the summer to study
housing. At this point, I was getting close to embarking on
my thesis, which would consume the sixth and final year. The
thesis idea I had initially pursued was a new design for the
Knesset, the parliament of Israel. I was intending to design
it not like the classical building that was actually built but as
something more organic and Middle Eastern. I spent a lot of
time doodling Knessets. In my imagination, it was a building
that would have a lot of domes—something with a romantic
Ottoman flavor reinterpreted in a contemporary way.

I applied for the scholarship and was selected as the
student from McGill. The program turned out to be a profoundly important experience. I met new people—my peers
in the program, who had been trained just like me but who
brought different perspectives. One of them, Michel Barcelo,
from the University of Montreal, was French Canadian, and
through him, I came to know better the rapidly energizing
movement known as separatism. Of even more consequence,
the intense confrontation with the realities of housing in the
United States and Canada—high-rise, low-rise, urban, suburban,
dense, sprawling—led me to abandon the Knesset idea as a
thesis project and revise my ambitions in a life-changing way.

Five of us participated, along with an accompanying faculty member. We traveled the continent. We visited Fallingwater. We met with Willo von Moltke, a former city planner in
Philadelphia, and with Ed Bacon, then the current city planner
of Philadelphia and the man responsible for Penn's Landing

and Independence Mall. I remember being deeply impressed with Louis Kahn's Richards building, a laboratory on the University of Pennsylvania campus, which we visited. But more to the point of the fellowship, we toured high-rise public-housing projects all over the United States and Canada. We drove the winding streets of the endless tract housing that rings every city. We visited Levittown and Mies van der Rohe's luxury high-rise apartments in Detroit and Chicago. All told, it was an extraordinary boots-on-the-ground survey of mid-century American urbanism and housing.

The Cabrini-Green complex in Chicago—typical of the depressing public-housing projects encountered during my fellowship.

I was devastated by the public-housing projects that we saw. They were monotonous, brick-clad, extruded towers with small windows. The exterior access walkways were enclosed by chain-link fencing, as if the residents were being confined in cages. The buildings were typically set in a sea of asphalt. Touring these projects was soul crushing. And yet I also registered that, in terms of design, they were not much different from luxury high-rise apartments: the same elevators, the same corridors, the same relationship, or lack of it, to the ground below. The luxury locations were better, and the finishes were fancier—but the architectural DNA was the same. I remember walking through Mies's Lafayette Park, in Detroit, and seeing firsthand how a design aesthetic had been imposed in conflict with people's actual lifestyle. The buildings were all glass, both the towers and the lower town houses. Walking in front of the town houses, you could look straight in. People had actually hung curtains behind the plate glass, with a smaller square window cut out of them. There were no balconies or terraces. There was no access from the units to the outdoors. It was relentless: the "universal" steel frame and impenetrable glass envelope. This is not how people want to live, I thought.

Mies's Barcelona Pavilion, designed for the 1929 world's fair, is a masterpiece, a great work of architecture. But how his vision evolved into high-rise apartments and office buildings left a lot to be desired. In the Barcelona Pavilion, Mies sculpted a structure whose planes and columns create a kind of open network that extends wondrously into the space surrounding it. But when it came to apartments and office buildings, he opted for elegant but simplistic containers with undifferentiated floor plates, stacked one atop another—as if that would create an environment in which human beings might like to live or work.

On this trip I was also struck by how wedded many Americans had become, if they could afford it, to buying houses in the suburbs. They were escaping the stress and strife of the central cities, but above all they wanted their own single-family house with its garden, its fence, its zone of privacy. What they wanted was the opposite of life in urban towers.

But there was also an in-between to discover, though one had to look for it. I'm thinking, for instance, of places like Chatham Village, designed by Clarence Stein and Henry Wright, in Pittsburgh, where town houses with pedestrian paths and clusters of parking offered a way of life more compact than any suburb but with many of the features that suburbanites wanted—the greenery, the privacy, a better balance between personal and community life.

By the time I returned to McGill, I knew what I wanted my thesis to be. I wanted to describe why suburbs are seen as desirable but also what they fail to achieve; and why apartment buildings are hated, even as they seek to resolve the inescapable challenge of urban density. The challenge I posed was how to provide the quality of life of a suburban house with its garden but do so in a high-rise structure. I began to think not of a specific project for a specific site but of a "housing system"— an approach to housing that could be adapted anywhere and that embodied certain core principles. These principles had to do not only with the design of the individual residential units but also with how they might be built. The thesis I wrote was called "A Case for City Living"—and one of the objectives was encapsulated as "For Everyone a Garden."

The obvious choice for a thesis adviser would have been Douglas Shadbolt, because he was a full-time member of the faculty. Instead, I sought out an architect named Sandy van Ginkel, whom I had been fortunate to hear at a lecture one day. He was not on the faculty, just a visitor. Van Ginkel had worked with the influential Dutch architect Aldo van Eyck and was part of a group known as Team 10, whose members aimed their barbs at dogmatic modernists, not because they opposed modernism but because they hoped to take it

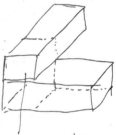

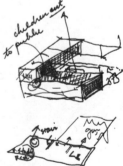

Work in progress, 1960: early sketches as I experimented with ideas for my thesis.

a step further. The group included Louis Kahn, Giancarlo De Carlo, and Georges Candilis, and was all about bringing the sensibilities of indigenous architecture and traditional urbanism back into the modern movement. Van Eyck referred memorably to his quest for a *casbah organisée*, evoking the dense vitality of the casbah, or old city, of urban areas of the Middle East and North Africa. These architects wanted to move away from the austere vision of pristine towers in a pristine park and expand concepts for urban development that better represented the complexity of actual life.

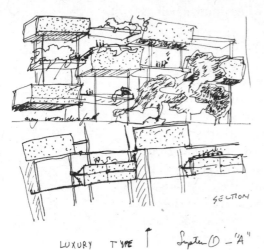

LUXURY   TYPE

Another preliminary sketch as "A Case for City Living" took shape.

Van Ginkel was a character— eccentric, worldly, sophisticated, and intimately connected to the European architectural world. He had been a member of the Dutch resistance during the Second World War. His brownish-blond hair and mustache softened a squarish head, and he spoke meticulous English with the pleasing inflections of his native land. He wore beautiful silk ties, prompting me to wear ties myself for a while. He loved music, Bach especially, and in conversation would bring up the fugal structure of Bach's musical offerings, relating it to the rhythmic structures of architecture. We talked about painting—I had become interested in Cubism, finding its spatial implications a source of excitement. I would talk with van Ginkel about what I was reading. I was fascinated at the time by *Architecture without Architects*, by Bernard Rudofsky, a book that explored vernacular architecture—architecture created by ordinary people, as you find in Italian hill towns and at the core of Middle Eastern towns and cities: the world of bazaars, courtyards, and terraced buildings following the shape of the land. One realized, for example, that the streetscape of ancient Athens would not have looked like a collection of mini Parthenons. It would have looked more like the clustered dwellings you find on Mykonos or Santorini to this day.

I would go to van Ginkel's house every week, bringing with me whatever I had developed since our last meeting: large sheets of tracing paper covered with sketches and careful notes—a diary of ideas. I still work like this, using hardbound sketchbooks, but back then my sketch pads were large and unwieldy. When I carried them outdoors, the Montreal wind made them billow like sails. I also brought models of my work, painstakingly crafted from cardboard or wood blocks. This was long before 3D printing made the building of models easier. I had to cut out every tiny window myself with an X-Acto knife, which sometimes slipped. Those early models bore streaks of red.

A 1962 sketch on the back of a business card—a precursor version of Habitat.

What I called a "housing system" in my thesis was actually three systems: variations on a theme but with differing modes of construction and different densities.

The first system was applied to tall structures, buildings from twenty to forty stories high. It consisted of a framed structure into which "boxes," representing the living units, were inserted. Each rested on the one below, in spiral formation, thus enabling the roof of one unit to support a slice of garden for the unit directly above. The second system was similar but applied to structures that were not so tall. The boxes were piled up in similar fashion, giving each unit a garden atop the roof of a lower unit, but there was no frame to bear the load—the units could support one another. The third system consisted of prefabricated walls that shifted at ninety-degree angles from floor to floor and were set back to allow open gardens for each unit, the units themselves being somewhat akin to town houses.

At work on my thesis, in both two and three dimensions. Professor Stuart Wilson is in the background.

There was a certain amount of mystification in the school about what I was doing—about what a "system" was. Everybody else was designing museums and opera houses and libraries for their theses, an impulse I certainly understood. I had myself spent months turning out Knessets. My classmates were designing specific buildings for specific sites. One person who was not mystified by my specific ambition was Jane Drew, a prominent British architect who, with Le Corbusier, had planned the Indian city of Chandigarh. She had come for a visit to McGill and was brought to the studio to review the work being done by sixth-year students. She came over to see my models, and I walked her through the concept. She liked what she was seeing and hearing. The school itself started paying closer attention. That was a turning point. But a "housing system" was still a hard sell in most quarters. McGill submitted my thesis that year as its nominee for a special prize for best academic thesis—every Canadian school could submit one candidate—and it must have landed with a thud. The prize was won that year by a design for an opera house in Toronto. I realize today how radical my thesis must have appeared.

* * *

I graduated from McGill in 1961. My years there had been a success. They had also been exhausting. Architecture is labor-intensive and time-intensive, and the intensity begins in school. Architecture students work around the clock when they're involved in a design presentation—known as a charette, derived from the Beaux-Arts days, when a *charrette*, or carriage, would arrive late at night to fetch the drawings to bring to a jury. Many sleepless nights and caffeinated days precede the moment when drawings are pinned to the wall.

I also had a family to consider. Nina and I had married in 1959, during my fifth year of school, and our daughter, Taal, was born while I was working on my thesis. I was young and had never experienced what you might call a normal bachelorhood—my relationship with Nina was the first I had ever had. We rented an apartment near McGill, and Nina got a job as a clerk with the Bell Telephone Company. I enjoyed fatherhood. Though I was steeped in work, I tried always to make time for the playground or, in those long winters, for sledding in the snow. I look back in astonishment to think that I began raising a family when I was twenty-two—fairly typical at the time—whereas my own children did not begin raising their own until they were in their thirties.

We lived frugally, with my parents sometimes helping out. The Bronfman girls—heirs to the Seagram's fortune—had attended the same high school I did, and I remember my father once observing, while we were driving, that it's just as easy to fall in love with a rich woman. This was before I married. I noted with quiet amusement that he himself had not done so.

What I knew for certain when I graduated was that I could find professional fulfillment in architecture, and already had. I had even begun to be recognized. At graduation, I was awarded the gold medal and other prizes, as well as the traveling scholarship. My parents were proud. My father was appeased. With the help of van Ginkel, my thesis was published in the important Dutch architectural magazine *Forum*. Aldo van Eyck was its editor.

The Richards Medical Research Laboratories, designed by Louis Kahn and completed in 1960, Philadelphia, Pennsylvania.

I went to work for Sandy van Ginkel for a year at his office in Westmount, a small, five-person studio in the basement of his town house. I was eager to get experience with actual construction, but as it happened, I spent most of that year developing a plan for a new town outside Toronto—Meadowvale—and not on individual buildings. The process began as it had with my thesis—making sketches and models, all of which were then reviewed with Sandy. His criticism and suggestions gave me the tools of an urbanist and made me aware that an architect cannot design a building in a vacuum, in isolation, which for architects can be a natural tendency and temptation. Henceforward, I was always aware that a building was part of a whole—the surrounding city, the neighborhood, the landscape. The sum total was as significant as the individual building being designed. I have not had any formal training in urban design but have always thought of that year at van Ginkel's as my PhD.

To know Sandy van Ginkel was to become more sophisticated—not in a smug way but in an organic way that had enduring value. I remember taking all of my savings, fifty dollars at the time, and buying a pre-Columbian bowl to give to him as a token of appreciation for everything he had taught me.

Louis Kahn,
c. 1970.

Meanwhile, my thoughts kept drifting to Louis Kahn—
we had visited the Richards Medical Research Laboratories
when I had the traveling scholarship, and I had admired
his work. I had the temerity to entertain the thought that
his office might be a place where I could get the building
experience I wanted. I decided to see whether I could secure
an appointment. Sandy's wife, Blanche Lemco van Ginkel,
had been a professor of urbanism at Penn and knew Kahn
well. She made the introduction.

I knew from study and observation that Kahn brought a
kind of compassion to the building process, a sensitivity to
the material language he was working with. Seeing the towers
and stacked laboratories of the Richards building, on the Penn
campus, I realized that something radical was at play. Kahn
understood that the mechanical shafts of a laboratory could
be an architectural element in their own right—they were
towers to be displayed rather than chimneys to be concealed.
He spoke of "served" and "servant" spaces. *Served* spaces were
the primary spaces that people actually used; *servant* spaces
accommodated stairs, storage, and the complicated apparatus,
such as heating and cooling systems, that made the served
spaces usable. The buildings of the past were singular systems
of construction—masonry, or brick, or wood, or wood and
masonry combined. They were about structure and envelope,
period. The new buildings were about multiple systems—
wiring, elevators, air-conditioning. Ducts were needed to
bring air in and take air out, in tremendous volumes. With
Kahn, for the first time, somebody was saying: We don't have
to sneak these systems in—we can make an architecture that
gives all these systems their rightful expression. We can allow
them to be seen. In his words, "Let the building be what it
wants to be." And architecture would be richer for it.

Kahn was saying something fresh and original that had
far-reaching implications. Most people don't realize the extent
to which high-tech practitioners like Richard Rogers, Renzo
Piano, and Norman Foster derived their language from Kahn's
philosophy. Consider the Centre Pompidou in Paris (Rogers
and Piano), the Lloyd's headquarters in London (Rogers), and
the Hongkong and Shanghai Banking Corporation (HSBC)
building in Hong Kong (Foster). But his influence has filtered
everywhere: rehabbed lofts in Tribeca, rehabbed factories in
Pittsburgh, my own rehabbed design studios in Somerville.

In all of these, the muscles and sinews of the buildings are at once visible and beautiful.

Physically, Kahn was badly scarred from a burn in childhood. The skin on his face was distorted and in places seemed melted. He was short and stout, with a shock of white hair and penetrating blue eyes. He could be charming—he had a stream of lovers—and he was intense. He was also a poet, who knew how to frame and articulate his vision. When he wanted to make a point about materiality and the essential impact of materials on architecture, he recited a little parable about what a brick wants to be: "If you think of Brick, you say to Brick, 'What do you want, Brick?' And Brick says to you, 'I like an Arch.' And if you say to Brick, 'Look, arches are expensive, and I can use a concrete lintel over you. What do you think of that, Brick?' Brick says, 'I like an Arch.' And it's important, you see, that you honor the material that you use. . . . You can only do it if you honor the brick and glorify the brick instead of shortchanging it."

The idea, in essence, is that certain forms and construction methods are inherent in the building materials themselves. But the idea expands to encompass not just what the brick wants to be but also what the building wants to be.

Detail from Kahn's desk calendar, April 1962, noting our appointment— a chance discovery in the Kahn archives.

Blanche Lemco van Ginkel's introduction must have worked its magic: Kahn agreed to see me. I went down to Philadelphia in April 1962, and Kahn met me in his own office. It was a Sunday morning, but a few people were on the premises working. Kahn was about to have cataract surgery. The lenses of his glasses seemed impossibly thick. I had brought my thesis drawings with me. We opened them up, and he paged through them carefully, turning the sheets one at a time. And then he said, "OK. I'm going to have surgery. You can start in September." I was elated.

A September starting date left the summer free. I had been awarded another traveling scholarship at graduation, and I needed to use it. Nina and I bought a Peugeot. We left Taal, who was just a year old, with my parents and set off on a two-month trip, driving across the northern United States, down the coast of California, then on to Mexico City and the Yucatán, back along the Gulf Coast and the Deep South, and eventually home to Montreal. Driving a Peugeot—a 1959 Model 403—was, of course, a mistake. It broke down frequently, and in the United States parts were scarce. Americans didn't even have metric tools. On our way back we met a Ku Klux Klansman who owned the gas station in Alabama where our car had found refuge after its most serious breakdown. He didn't like Jews, but he liked Israelis—he thought Israelis were somehow different.

Nina and I moved to Philadelphia in September. When I arrived at Kahn's office, I was twenty-three, the youngest person there, and the only one who had not been Kahn's student or known him for many years. I plunged in—there was a lot going on. Kahn was designing the Mikveh Israel Synagogue, in Philadelphia, a project that in the end never got built. He had won a commission in India to design the Indian Institute of Management, in Ahmedabad, which did indeed get built. And he was deeply involved in the design of the Salk Institute for Biological Studies, in La Jolla, California, which would become one of his signature achievements.

Kahn employed about twenty-five people. Two of them became my close friends. One was Anne Tyng, Kahn's lover and the mother of his daughter Alex. Tyng was a brilliant architect in her own right, credited with bringing Kahn into a series of projects that explored the geometry of Platonic solids and space frames. The other, David Rinehart, was a

A 1963 sketch by Kahn of the proposed Mikveh Israel Synagogue, which was to have been built in Philadelphia.

few years older than me and a graduate of Penn—a gifted designer with a beautiful natural sketching hand.

Somehow, I found myself given a lot of responsibility. I started working on a competition that Kahn and Anne Tyng were doing for a museum at the University of California, Berkeley. Then Kahn assigned me to work on the synagogue. The fact that I was an Israeli may have had something to do with his decision. I was invited to join meetings with the client, and worked closely with Alan Levy, who was Kahn's project manager on the job. It was an extraordinary experience.

My own drawing for the Mikveh Israel project, made while I was apprenticing with Kahn, 1963.

In an office so small, one absorbs everything. I could join meetings with engineers and discuss building details with far more experienced architects. I learned to draw more precisely and elegantly than I ever had before. The precision required in architecture—the need to ensure that drawings and numbers and notations are clearly understood, but also that presentations look beautiful to clients—is one reason, historically, that the handwriting of architects has been elegant. Whether this will survive the age of the computer is anyone's guess.

Working with Kahn, even the small things sometimes turned out to be more significant than they might seem. I watched him use charcoal for sketching, essentially as a tool of thinking. His fingers were an extension of his brain. The marvelous thing about charcoal is that you can draw a plan or a section, think about it, and if you don't like what you see you can just wipe it, or a part of it, away and try again. You draw a circle, and if it doesn't look quite right, you pass your hand over and draw an ellipse. Charcoal is malleable. That fluidity of thinking—the organic connection of eye, hand, and brain, which generates its own special form of creativity—is something that I worry that younger architects today, in the computer age, may be missing. Computers produce astonishing imagery and allow for endless experimentation. They are essential. But you can become a prisoner of the computer, and you can fall in love with a design because a computer renders it superbly. On a computer, the scribbled forms and notes and smudges of a sketchbook page turn into imagery that might have burst from the studio of George Lucas's Industrial Light & Magic.

At Kahn's office, before computers, we sometimes used to joke, in an admiring sort of way, about Romaldo Giurgola, an eminent architect, close to Kahn, who had a golden hand. He drew so well that anything looked good. The light, the shadows, the trees, the glints on the windows, the figures strolling outside: it was all perfect. You could almost feel the soft breeze of a summer afternoon and breathe in the scent of jasmine. A rendering that is too perfect can be deceptive—it can trick one into thinking that the reality suggested by the image is better than it is.

Kahn was methodically, systematically engaged in every phase of the architectural endeavor: The charcoal phase—embryonic, giving form. The pen phase—working things out in great detail. The construction phase—putting materials together, learning from the mock-ups, changing designs as a result. And after that, he was involved personally in the building process—down to the smallest detail—making decisions along the way.

* * *

A happy and productive year went by in Kahn's office. On my own time, I was also engaged in other projects outside Kahn's office if not outside his shadow. A competition had been announced for redeveloping the center of Tel Aviv. Kahn was on the jury, so no one in the firm could participate. But David, Anne, and I decided to develop a scheme for Tel Aviv just as a theoretical exercise, even though we could not submit it. We were driven by ideas of how a new city might grow organically and yet rationally—*gnomonic* growth, as it's called, like that of a nautilus shell, which grows and enlarges without replacing the earliest portions of its constructed existence. I was also keen on further advancing the ideas elaborated in my thesis—making them into something real.

In other spare moments, and with my thesis in mind, I started designing a housing scheme for Palestinian refugees in a city of my imagination to be built in Egypt, near the pyramids of Giza. In retrospect, it seems an exercise in sheer fantasy, and I recognize too the well-meaning clumsiness of a distant foreigner who presumes to have the answers. But I

An exercise in
naivete: my 1963
plan for a city to
house Palestinian
refugees, to
be built near
the pyramids
in Giza, Egypt.

was serious about the idea and the possibility of advancing
it. I produced some plans and sent them to Prince Sadrud-
din Aga Khan, who was then the High Commissioner for
Refugees at the United Nations. I also wrote, requesting
an appointment. I don't recall getting a response, but I still
have the drawings.

Recently I looked through some material in the Kahn
archives that I had not seen in years—and in one case had
never seen. I realized—I had forgotten—that I had designed
the gutters for the Salk Institute. I had also suggested adjusting the orientation of the Indian Institute of Management
to provide more shade and to face the prevailing winds—and
the suggestion had been accepted. One item from the Kahn
archives was Kahn's personal desk calendar, with the name
"Moshe Safdie" written in on the day of my first interview.
Another item was the original letter of introduction that
Blanche Lemco van Ginkel had written.

The friendship between David Rinehart and Nina and me
had been memorable and deep. He was a part of our household. Among other things, I was hoping to go to India with
David to oversee the construction of the Indian Institute of
Management, bringing Nina and Taal for the duration. But it
was not to be. Sandy van Ginkel showed up in Philadelphia
one day in early 1963, coming back into my life and bearing
a piece of news: Montreal was going to host a world's fair.
The target date was 1967. Van Ginkel would be in charge of
preparing the master plan.

He asked, would I come back to Montreal and head the
team responsible for designing it?

Beyond the half-century mark: Habitat '67, in Montreal, Canada, as it is today.

CHAPTER THREE

The World of Habitat

Under Louis Kahn's care, I had become immersed in a world that had everything to do with the conception of buildings and also with the *craft* of building. By *craft* I mean how, beyond the spatial imagining of what a building is going to look like, you conceive and work out its construction—thinking hard about the materials, the fabrication, the sheer making of a structure. Without this immersion, short as it lasted, I could never have led an inexperienced team of young architects to realize Habitat '67, now recognized as one of the most demanding and technically innovative building projects of its time.

Kahn had not been particularly happy to see me leave Philadelphia. He was also not greatly pleased, six months after my departure, when I invited August Komendant to come work with me on Habitat '67. Komendant served as the structural designer for most of Louis Kahn's masterly concrete projects: the Richards building at Penn, the Salk Institute in La Jolla, and later the Kimbell Art Museum, in Fort Worth, Texas.

In August 1963, our family took an apartment near McGill, in the area around the university that was referred as the student ghetto. A neighbor down the street was Leonard Cohen, fellow Westmount alumnus. The writer Mordecai Richler lived on the other side of campus. Montreal was booming, and there was great excitement about Expo 67—no one called it the 1967 International and Universal Exposition, its official name. There was also considerable excitement that this was a celebration of Canada's centennial as a nation. It was a twilight moment, just before separatist sentiment began to shake the politics of Quebec in earnest, though ferment in the French community was already palpable. The mayor of Montreal, Jean Drapeau, was a leader of the French cause and a major force in the city's development. Balding, with a dark mustache and heavy, dark-rimmed glasses, he cultivated the aura of a populist, as indeed he was, though he had a taste for the good things in life. "A cross between Walt Disney and Al Capone" is the description of him that everyone remembers. Montreal was a bicultural and bilingual city, but at the time a largely unmixed one, with the English in the west and the French in the east. It was still Canada's financial and corporate center—functions that have now mostly shifted to Toronto.

The early 1960s were an exciting time to live in Montreal, and the planning for Expo seemed to involve everyone and to touch everything. Architecturally, Montreal became something like a watering hole in the jungle, a place where all species safely congregate, as prominent architects from around the world—Frei Otto from Germany, Buckminster Fuller from the United States—arrived to design buildings for the fair.

In my new role, I had two responsibilities. The first was working with Sandy van Ginkel and the team on the master plan for Expo. This would be the overarching scheme for the world's fair—the arrangement and preparation of the site itself, the layout of pavilions on the site, the traffic- and pedestrian-circulation patterns, the degree of integration with Montreal. The second responsibility, which was a subset of the first, was developing the ideas that had percolated in my thesis, and that as yet existed only on paper, into an actual project—something not only built but also animated by human life and activity. This second task, which led directly to Habitat, was not master planning; it involved the creation of a structure or group of structures that would be the equivalent, in effect, of one of the many pavilions at the world's fair, though with a purpose and ambition entirely different from those of the others. The team that van Ginkel put together for the Expo master plan consisted of myself, effectively as team leader; Adèle Naudé (later Santos), who would go on to become the dean of architecture at MIT and a very important educator and practitioner; Jerry Miller, who had trained at McGill and Harvard; the architect-urbanist Steven Staples; and a handful of others.

While Expo resources were made available for what was to become the Habitat effort, I realized we would need substantial additional resources to put it together. Salvation came in the form of the Montreal architect Jean-Louis Lalonde, who was hovering around the Expo offices as a representative of Canada's cement companies, and who had $50,000 available as a fund to encourage the use of concrete at the world's fair. I had not yet concluded that Habitat should be built of concrete, but I jumped at the opportunity and accepted the grant, with no commitment in advance that concrete would actually be used (though, as things turned out, it would be). The grant seems small when viewed through a contemporary

My mentor in Montreal, Sandy van Ginkel, in the 1960s.

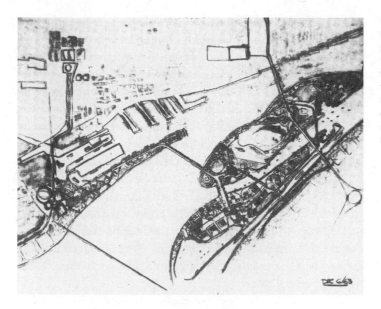

My sketch for the Expo 67 master plan, December 1963. The world's fair would occupy the two islands. Habitat was designed for the narrow peninsula by the port.

lens, but it made all the difference and gave me the resources to create the nucleus of the Habitat team. I put together a small group, among them David Rinehart, who joined me from Philadelphia. I also invited August Komendant, and as soon as he saw our study models, with the many little blocks piled up in many possible formations, he lit up and accepted the assignment—drawn by the concept itself and by the engineering challenge posed by what we had in mind.

The master-plan effort necessarily took precedence over the Habitat effort. Mayor Drapeau had conceived the idea that Expo would be accommodated on two islands in the Saint Lawrence River across from the port and downtown Montreal. One of the islands already existed but would have to be expanded, and the other would have to be created from scratch. Subway links and vehicular bridges would need to be built to connect the islands to the city.

Van Ginkel and the team had another agenda—to use Expo as leverage for revitalizing the Montreal riverfront on the city side of the Saint Lawrence. This involved extending Expo from the islands to the mainland—to Mackay Pier, later renamed Cité du Havre, a narrow ribbon of peninsula within the port area whose development, we believed, would help bring active city life back to the industrial waterfront. The head of the port authority objected, wanting no change to his gritty fiefdom, but he was overruled. Locating Habitat—a dynamic mixed-use urban precinct—on this site became part of the larger vision. It would be linked to Expo—and Expo linked to the city as a whole—by a novel transportation scheme that had three separate circulatory systems, all synchronized to serve different functions, different destinations, and different speeds, but linked to one another at various points.

To expand one island and create another, trucks brought in rock and dredging fill around the clock. The Habitat site, originally not much more than a rocky spine protecting the port, needed to be enlarged and turned into something one could actually build on. Meanwhile, an upstream barrier, or "ice bridge," was being installed across the Saint Lawrence to break up ice as it flowed downriver. Ice floes and icebergs could easily damage the islands and other created land or even sweep the land away altogether. The ice bridge is still needed and still there.

Being a joint venture between three levels of government— the federal, the provincial, and the municipality of Montreal— Expo was a complex affair. The federal government initially provided 50 percent of the funding for the world's fair, the provincial government 37.5 percent, and the city 12.5 percent. Under the jurisdiction of the ministry of trade and commerce, a triumvirate was appointed, drawing on establishment and political figures to make the master plan a reality. There was a lot of bickering and posturing, a lot of indecision, and a lot of wasted time. Eventually, this first triumvirate was replaced by a second one, consisting of people who knew how to roll up their sleeves and get things done. One member of this second triumvirate was Pierre Dupuy, Canada's ambassador to France. Dupuy had experience negotiating with other countries—something a world's fair needed. Another member was Robert Shaw, president of the Foundation Company of Canada, a major contractor. The third was Colonel Ed Churchill, director of the corps of engineers of the Canadian army. Shaw and Churchill had just completed the construction of the so-called DEW line in the Canadian Arctic—the "distant early warning" system against Soviet missile attack. This second triumvirate, which had overall authority but answered to the federal cabinet, brought Expo and Habitat to fruition.

The original concept for Expo 67—meant to distinguish it from any previous world's fair—was that it would be organized heavily around themes rather than simply around the individual pavilions that countries and companies might build. Man and His World was the overarching concept, and there were subthemes such as Man in the Community and Man in Nature. Countries were encouraged not to emphasize stand-alone pavilions as expressions of national pride but to create exhibits that could be incorporated into pavilions devoted to the various themes. Also contrary to previous fairs, Expo would not be symbolized primarily by some pharaonic physical monument—a tower or spire. It would be most powerfully symbolized by ideas. This had all been hammered out at a blue-ribbon conference at the Château Montebello resort, where representatives from every imaginable constituency in Canada had gathered to discuss the vision for Expo. Except for the use of the word "Man," the original Expo concept stands up well even after the passage of a half century.

As reality played out, though, the concept ran into obstacles. Countries insisted on their own pavilions, and these would be duly built. The themes never achieved quite the international pride of place for which the planners had hoped but were mostly Canadian contributions. And Jean Drapeau, the mayor, was adamant about having an imposing physical monument to memorialize the fair.

In the fall of 1963, as our team was in full swing finalizing the master plan, Pierre Dupuy, perhaps with a nod from Mayor Drapeau, invited a Beaux-Arts-style Parisian architect, Eugene Beaudoin, to create an alternative master plan to submit to the triumvirate. What Dupuy was thinking remains unclear. Beaudoin came up with a concept that involved abandoning any connection to the mainland and focusing on one island in the river, shaping it like a ship, with an imposing forty-story-tall spire at the prow. Those of us in the Expo planning group were appalled at the kitschiness of the idea and decided to fight back. We hired a young lawyer, Claude-Armand Sheppard, an aggressive rising star whom I'd known at McGill, and went public with our concerns. What this got us was a fresh, engaged hearing by the reigning triumvirate, particularly Shaw and Churchill, who invited us to present our plan in detail and discuss its ideas as compared to Beaudoin's. I remember a lively discussion, with others from Expo's management present. Eventually, Beaudoin was sent home. His scheme was regarded by the triumvirate as not serious; it failed to address holistically the full range of issues involved.

Drapeau, though, had long been entranced by the notion of a tower. He had once entertained the idea of dismantling the Eiffel Tower and having it reassembled in Montreal for

the duration of the fair. At one point, in a city-hall ceremony, he unveiled his own plan for a tower that would have been 1,967 feet tall—a proposal that collapsed of its own financial weight. It was clear from early on that the choice was between Habitat '67 and a Drapeau tower as a potent symbol for Expo, and the tower's ultimate failure was not something the mayor ever forgot. Several years later, when Montreal hosted the 1976 Olympics, a project firmly under his control, Drapeau kept me at arm's length.

With Beaudoin back in Paris, we proceeded to finalize the master plan. To be sure, there would be many developments and changes to come. At one point, it was calculated that the dredging boats in the Saint Lawrence would never get enough sand and fill to form the islands as planned. Panic ensued, as this appeared to threaten the opening of the fair. The planning team came forward with a novel idea that turned the crisis into an opportunity. We suggested creating ponds, canals, and lakes within the islands, thus reducing the need for fill. These water features became one of the most compelling aspects of the fair as part of a landscaped public realm.

In 1964, with the master plan finally on track, it was time to turn our attention to Habitat. We were operating on a shoestring. Expo didn't have much of a budget to underwrite a feasibility study, much less a budget to cover the cost of the project itself. But Jean-Louis Lalonde had stepped in with a grant. He also provided something else: the name Habitat, noting that the word connotes housing in both English and French.

The original idea behind Habitat was to create an urban sector, an entire community occupying much of the enlarged Mackay Pier. In my thesis a few years earlier, I had envisaged housing modules stacked twenty to thirty stories high in a frame-like tower structure, but now I was thinking that if one could get the arrangement of units to lean back as if on a hillside—slopes of stacked houses, stepping back floor by floor—we could have gardens and other areas open to sky, sun, and rain. The "hillsides" would hover over sheltered public spaces on the ground below and would be laced with "streets" every four floors to provide access. A-frame structures would be needed to support and stabilize the hillsides. We placed inclined elevators into the A-frames for vertical circulation and won a promise from the Otis Elevator Company to produce them. Toward the end of the pier, the land got too narrow to accommodate the A-frames, so we designed a smaller, twelve-story-high, village-like cluster for the tip. All in all, we envisaged a Habitat with twelve hundred dwellings and a variety of community facilities, including schools, shopping, and office space. The planning for Habitat was hammered out with David Rinehart during countless late-night sessions. Komendant played a crucial role in all this, devising engineering solutions so that we could be sure that what we wanted to achieve was consistent with the laws of physics.

The Habitat concept was informed by merging two sets of ideas about private living space and community life—ideas that, as I now understand, owe much to what I saw and experienced when I was growing up. Individual units, whether houses or apartments, needed to be well-defined, with a strong sense of physical identity. They needed to have compelling views in more than one direction. They needed access to private outdoor space, such as gardens, as well as to communal outdoor space. The terraces had to support significant plant life—not sorry shrubs, as in so many developments, but something closer to nature. The community

Model of the original plan for Habitat '67, 1964, before budget realities dictated a scaling back of the vision.

had to be safe—children should be able to wander by themselves within the clusters. Gardens, pedestrian pathways, and common rooms should encourage residents to interact with one another—indeed, should make a certain amount of casual interaction unavoidable. On the ground, under the residential units and partially open to the sky, with a view of the river and the harbor, was the public hub: the schools, the stores, the day-care facilities—a network of active urban spaces separated from vehicular traffic.

All our thinking about Habitat—the overarching ideas and the physical manifestation of them—was eventually embodied in a large and detailed physical model, some five feet by nine feet in size. I could not help thinking of the model I had made in plaster with my friend at the Reali School—it was about the same size but a lot lighter. What we needed now was an indication of government support. I made a pilgrimage to Ottawa to visit the Central Mortgage and Housing Corporation—Canada's equivalent of the U.S. Department of Housing and Urban Development. It was an agency I knew well because it had sponsored one of my traveling fellowships. At the time, I had also worked for a spell in its offices. The corporation was headed by Ian Maclennan, an architect and a dynamic institutional leader—the sort of person who rarely makes headlines but knows how to drive a bureaucracy toward the public good. He was excited about the Habitat project and broached an idea that might, he believed, give us the wherewithal to make Habitat a reality: connect with a developer. The developer he had in mind was William Zeckendorf.

As a first step, he suggested that we retain the services of two of Zeckendorf's associates—Stewart M. "Bud" Andrews and Eric Bell—to do an in-depth feasibility study and provide the kind of detailed information that would be needed by financiers. Zeckendorf at the time was one of the most significant developers in North America. He owned the Chrysler Building, in Manhattan. He had built Place Ville-Marie, in Montreal, designed by Harry Cobb and I. M. Pei, who had become Zeckendorf's court architects. Zeckendorf was a larger-than-life figure, who served lobster lunches on a private jet as he traveled like a pasha from site to site. Andrews and Bell joined the team to see whether they could arrange a marriage between our Habitat vision and financial reality. And they did. The consultants pegged the cost of the project at $42 million, and they also devised an ingenious financial strategy: if Zeckendorf were to be granted a federal tax break allowed to Canadian industry for the purposes of research, the financial boost would make the project commercially viable for private development. Under those conditions, Zeckendorf seemed willing to take the project on. The "research" idea made all the difference to the investors, but the claim was also true. Habitat was an innovative and experimental project whose ideas and methods, if successful, would have applications not only in Canada but worldwide.

Armed with Zeckendorf's interest and a workable financial plan, we pulled together all the documents and our very large Habitat model and prepared to present the idea for ultimate approval to the Canadian prime minister, Lester B. Pearson, and his cabinet in Ottawa. We flew from Montreal to Ottawa on a twin-engine chartered plane with the Expo triumvirate and a few others. In the Parliament of Canada building, we unpacked the model for Habitat. The prime minister arrived, the minister of finance, and other members of the government.

I opened with a general presentation—the blight of suburban sprawl, the need for greater density, the distaste of Canadian families for traditional apartments, the need to reinvent dense urban living. I spoke about the industrialization of the building process. The Habitat units were to be prefabricated, and in fact manufactured on-site. This was a new way of building on a large scale, offering many opportunities down the road for Canadian industries. The discussion seemed to be going well.

But then a warning light flashed. The minister of finance was concerned about the Zeckendorf tax break for research. He worried that it would set a dangerous precedent to have the tax break used this way—for a form of research that was arguably legitimate but could also be seen as special favor to a private developer—and that other developers might soon be lining up for similar treatment. One could see the furrowed brows as the politicians thought about precedent, risk, and optics. They withdrew to confer among themselves and finally determined that the tax-break proposal would not be considered. Zeckendorf was out. The better news was that the federal, provincial, and city governments would themselves be willing to step up and play a role, but only for

$15 million—this for a project budgeted at $42 million. Do what you can, we were told.

I was in shock for a day or so, convinced that the last chapter of the Habitat story had been written: the dream would remain a dream. Then I started asking myself: What might we do with $15 million? We could not build the project as initially envisaged—an urban sector across the entire pier. But we probably could detach one piece of it. We did some calculations: a precasting factory on the site would be in itself $6 million, leaving some $9 million for the actual building. We could probably realize a section of Habitat comprising some 150 to 200 housing units.

I spent the next week making sketches and experimenting with models, and eventually we came up with the scheme that you can see today along Mackay Pier, on the Saint Lawrence. In essence, we built the small, twelve-story village section of Habitat envisaged for the tip of the peninsula. There was a lesson in this, one that I've found myself remembering throughout my career: sudden duress doesn't have to mean the end of the story—sometimes it gives you the material for a different story.

* * *

Take a look at Habitat '67, and one can see that it is not the easiest thing in the world to draw. In 1963, the era of computer-assisted design, or CAD, was decades away. The era of 3D printers was a half century away. Everything needed to be done by hand. Back then, an architect's office was a place of T squares, scales, pens and pencils, erasers, compasses, slide rules, and endless rolls of paper. The work was labor-intensive. Drafting was continual and had to be meticulous. Having to erase and redraw an enormous document because of some careless error was painful. Any change in a building demanded yet another drawing or much erasing and redrawing. Curves, double curves, and three-dimensional curves made the work even more complicated. For many of our designs today, computer-assisted design is essential. But models are by no means obsolete. I am personally drawn to them for both their beauty and their utility.

As I began working on my thesis, models were my prime tool. Thinking that we should be clustering and piling up apartments in a high-rise structure in a new kind of way, I cut wood into small blocks, each more or less representing an individual dwelling in terms of volume, and started to explore different ways of piling them one on top of the other. In time, my process became more systematic. I cut the blocks so that they were uniformly twice as long as they were wide. I could stack them like domino blocks—above one another in spiral formations. In each case, the result was space for an open garden on the roof of the box below. Models became a potent tool.

The presentation of my thesis had consisted of a large, multilevel frame, somewhat like a scaffold, made with balsa sticks, into which modules were inserted in diverse patterns. Four years later, we were able to take advantage of a major civilizational advance: the invention of Lego, which enabled one to click together brick-like modules in a variety of configurations. Our Habitat team must have bought up every Lego set in Montreal. At the time, Lego made only one brick size, but with that as the standard module we created an extraordinary richness of possibilities, clustering them into groups and then into ever larger groups. However, the solid nature of Lego made it impossible to imagine the living spaces inside the blocks—that is, inside the actual residential modules— so these we constructed, at first by hand, out of cardboard, tedious work that left fingers lacerated. Eventually we wised up and ordered hundreds of cardboard sheets from a factory that were cut and creased in advance so they could be folded easily into identical hollow boxes and then stacked according to whatever scheme we had in mind.

In the more than five decades since, technology has revolutionized the way we conceive a structure. And yet models remain central to our work. Today in our workshop in Somerville we have seven model makers. A few of them studied architecture and found themselves developing a passion for model making. Tony DePace, who has been our chief model maker for thirty-five years, studied carpentry and began apprenticing in an architectural-model shop. Today, two model makers translate CAD drawings into a form that can be fed into a laser cutter or a 3D printer. To be sure, we still have saws and drills and paint and glue. We still have a vacuum-forming machine, which allows us to create molds and then shape transparent surfaces by pressing acrylic sheets over the mold to get the curved glass for the models of structures like the Peabody Essex Museum, in Salem, Massachusetts, or the toroidal dome over Jewel

Downsizing Habitat '67 was aided by the invention of Lego. Experimenting with new configurations became easier.

Changi, in Singapore. Our 3D printers can also quickly fabricate all the elements of the entire fifty-story new tower for Singapore's Marina Bay Sands: every balcony and window division and scores of various details. At the center of our office atrium, rising from the basement level, we have on display the twelve-foot-high model for the 12-million-square-foot Raffles City complex in Chongqing, China. It's a working model that has been cut, chopped, and changed as the design developed.

Many architectural offices today have ceased to make physical models. And to be sure, it is possible, in a brief period of time, to feed a computer with spatial and geometric data and immediately generate walk-throughs, in theory showing everything a physical model could show. The viewer can glide through hallways as if experiencing virtual reality, watching the scenery change outside as one glances through the windows. It can be hypnotic. And yet, I find this capability insufficient: I'm never able to get as good a feel for the space, its scale, and its relationship to what surrounds it, as I do from a physical model. Physical models do have a disadvantage: you generally look down at them from above, and you need to bring your eye down to ground level as well. In our own work, communicating with clients by means of fly-through virtual reality is not as effective as communicating by means of a physical model. People respond to the physicality of built space, whether it's the finished structure or the model that preceded it.

As for Habitat, many of the models we made still exist, in various configurations and sizes. Most of them, now in the custody of McGill University, are in a warehouse in Montreal. And as if to close the loop, browsing the internet yields scores of images of Habitat rendered in Lego by various hobbyists. Several years ago, the Lego company asked consumers which iconic structure should come next in its architectural series, joining the White House and Trevi Fountain and other monuments among the packages on sale in stores. Habitat was the number-one suggestion, though for some reason, unfortunately, Lego declined to produce a Habitat set. Maybe by way of compensation, the company does market one now for another of our projects, Marina Bay Sands.

* * *

With a downsized design for Habitat, we were once again in business, and when the dust settled, we ended up with a twelve-story complex consisting of three clusters. A total of 354 prefab modules were configured into 158 residences. Two flying streets in the air would provide access from the elevator cores. The revised design was all set to go.

The high of that moment will forever be associated in my heart with a deeply personal tragedy. Upon our return to Montreal, Nina and I had had a second child, a boy we named Dan. A few days after Habitat was approved by the cabinet, a distraught call from Nina made me race home. She had discovered Dan dead in his crib—a case, we would learn, of sudden infant death syndrome. He was three months old. An ambulance was at the scene, but there was nothing to be done. It was a shattering moment, and I will not try to describe it. The memory of that day would cast a long shadow. I had the escape, if you can call it that, of immersion in work. Nina carried a special burden, made more difficult by the fact that so much of what I was doing was in the public eye. There were continual demands. Some of them were social, and many involved publicity of a positive kind. But there was also controversy. It was a hard period. Happily, we would soon have another child— our son, Oren.

Around this time, Ed Churchill, a member of the triumvirate, called me into his office. He and I had by then become very close. He had a sense of humor—he called our fierce resistance to the Beaudoin plan "the palace revolt"—and had a disconcerting way of seeming to grin even when he was angry, with his curly head of reddish hair and roundish face. He wasn't angry on this occasion. He had his usual grin, a twinkle in his eye. He had welcome news: we had gotten

A model of Habitat '67 as it would be built.

Ed Churchill (top),
a member of
Expo 67's ruling
triumvirate,
and August
Komendant, the
fabled structural
engineer.

final approval on the budget. Then he said, with a straight face: "And you're fired."

He elaborated. Habitat was your concept, he explained. You shouldn't execute it as a civil servant—that is, as if you were a bureaucrat simply executing someone else's idea. You should go and open an independent office—set up an architectural firm—and build Habitat on your own. Expo will be your client. Churchill suggested that I seek collaborative support from a prominent Canadian architectural firm.

This was only one of many memorable Churchill encounters. Many months later, when we needed building permits, it became apparent that the city of Montreal did not have an engineering department that could match the sophistication of our own structural engineer, August Komendant, who was a marvel. A German-speaking Estonian, he possessed a Teutonic bearing that made one sit up and take notice. The thick lenses of his glasses enlarged and intensified his blue eyes. In the late stages of World War II, he had served as a consulting engineer for the U.S. military in Europe, working for General George S. Patton, among others. One celebrated story, recounted in a biography of Louis Kahn, involved a damaged bridge that Patton needed to cross immediately. Komendant analyzed the load capacity of the bridge from one end to the other and then painted a curving white line across the surface to indicate a path the tanks could safely follow. Komendant was a pioneer in the use of prestressed concrete. He had developed many novel theories about structure. He was utterly confident in the soundness of Habitat's engineering, and I trusted him completely.

Others were not so sure, and Habitat's configuration of stacked boxes made some traditionalists uneasy. The chairs of the engineering departments at both McGill and the University of Toronto reviewed the plans and produced a report stating, in essence, that if Habitat was built as designed, it would collapse. And if it didn't collapse on its own, then an earthquake would bring it down. Ed Churchill called me in and outlined where we stood. He was taken aback by the report. But he said, "Let's see what the world thinks of Komendant." I sat in his office as he called clients in Philadelphia, where Komendant had worked with Kahn. I also listened as Churchill spoke to someone who knew about Komendant's wartime work, and to someone who knew about Komendant's work in an earthquake zone. After three or four calls it was clear that Komendant was a unique talent. Churchill was satisfied.

"I am the sovereign!" he said dramatically. He meant only that in his official capacity, he stood for the federal government, and in this matter his rule was law. He placed the critical report in his drawer. As we came close to Expo's opening, I received the Massey Medal for Architecture, in recognition of the design and creation of Habitat '67. It was a fine moment, but I felt that Churchill should be the one to have the medal. Not long afterward, I walked into his office and presented it to him. "This belongs to *you*," I said. When Churchill died, in 1978, I discovered to my surprise that he had left me the medal in his will. I prevailed upon his son to keep it in the family.

* * *

As I looked for a partner to assist my nascent architectural operation, my first thought was to go to one of the major modernist firms—specifically, to John B. Parkin Associates, in Toronto, which was the Canadian equivalent of the global and multifaceted Skidmore, Owings & Merrill. John B. Parkin was the principal and handled the business side of the firm. The design chief, John C. Parkin—oddly, no relation—was a Mies van der Rohe disciple who had designed many modernist public buildings in Canada and had worked with Mies himself on his buildings in downtown Toronto.

My first team: the entire Habitat '67 staff at the Place Ville-Marie offices, Montreal, 1965.

Parkin was a big firm, very corporate, and I felt a little looked down on when I approached them. Perhaps I took away the wrong impression. There was certainly a difference in style. The Parkin representatives were well-dressed, formal, and socially polished. My team was more ragtag—working in shirtsleeves, and with first names used by everyone. We didn't have rigid hierarchies. The Parkin people said they would love to do Habitat with us. But, of course, they would need to control the project, and they insisted on doing the engineering themselves, in-house. I said, No, Komendant is my engineer, and he's not going anywhere. Parkin tried to go over my head, but Churchill and Shaw backed me up, and I started looking elsewhere, this time back in Montreal. The firm David, Barott, Boulva was eager to collaborate and had no desire to take control. They sent three senior people to my office, where they would work with us for the next three years.

I had negotiated a contract with Expo—meaning with Ed Churchill and his staff—for our fee: a total of $600,000. From this, all the office expenses—rent, salaries, professional insurance—would have to come, up through the completion of the project. At its peak, forty people were working on it, some of them associates from McGill or architects I had come to know in Philadelphia. I rented space in Place Ville-Marie. I hired a secretary, the redoubtable Mercedes Aupy, who had worked at the Expo offices. She was one of those people who truly knows how to run someone's life. I also hired an accountant. Within a few weeks, we were all putting in fourteen-hour days. It's the only time in my

David Rinehart (holding the paper) with me and others at Place Ville-Marie, 1964.

With our daughter,
Taal, as Habitat '67
rises, 1966.

life when I have been able to work on a single project for a
sustained period of time, with no distractions. The senior
architects sent over by David, Barott, Boulva—a generation
older than everyone else—provided an important incre-
ment of know-how and experience. Their technical savvy on
matters such as waterproofing and general detailing proved
indispensable.

As soon as the project was funded and became a reality—
albeit yet to be built—the critics and opponents came out
of the woodwork, among them Paul O. Trépanier. Though
an architect, he was the mayor of the small city of Granby,
Quebec, and also a former president of the Ordre des Architectes du Québec. Trépanier started writing about our mad
project—"a wildly insane example of criminal naïveté," to use
his exact words—and in 1965 called for a royal commission to
investigate why something so stupid was being built. Trépanier
made headlines for a while, but cooler heads recognized his
motivations—publicity, with an eye on higher office—and the
challenge faded away.

At around this same time, a French development company, working with some local architects, unveiled a scheme
and proposed it to Expo. The company announced that it
could build an experimental housing project—on the very
site we had chosen—for half the price. What made this turn
of events especially galling was that the French company was
one we had consulted about prefabrication. We had shared
with them our proprietary drawings, our engineering studies,
our cost estimates—everything. And then they went behind
our backs. This was the sort of trial balloon that should have

Drawing by Porges #1967
The New Yorker Magazine, Inc.

been popped immediately. But orders came down from government officials in Ottawa that the French plan, known as Y67 (the structures were shaped in a Y), needed to be taken seriously—for political reasons, we assumed. The result was a kind of bake-off, with the Habitat team and the French team each presenting a scheme to Expo's management, which had authority over the world's fair as a whole. The French company had prepared a thick, slick, leather-bound set of documents on expensive paper. We, for our part, had all the documentary backup (on regular paper) and also a detailed physical model. The French proposal was dismissed as inappropriate. Neither the design nor the mode of construction was deemed sufficiently innovative. A few months later, astonished Expo executives received a bill from the French company to reimburse it for its work.

More dangers lay ahead, including a cabinet shuffle in the national government. The ministry of trade and commerce had ultimate jurisdiction over Expo as a whole, and the new minister, Robert Winters, voiced concern that the world's fair was running over budget. Winters, too, had ambitions for higher office and was looking for opportunities to position himself as a flinty watchdog. He eventually decided that Habitat should simply be canceled. By this time, January 1966, we had already produced about forty of the precast concrete units. The ministry commissioned a study to see whether the units could simply be dumped into the Saint Lawrence River—that is, to determine whether the river was deep enough to take them. In the end, the Winters proposal was itself dumped. We were too far along, and Expo's management as well as wiser voices in the government were

committed to Habitat. Ultimately, though, we did run into
scheduling problems. There were labor shortages—the entire
Expo site was a frenetic work zone, with ninety separate pavilions under construction. To ease the pressure, it was decided
we would execute the assembly of the modules so Habitat
would appear complete during Expo, but we would finish the
interiors of only two-thirds of the units, leaving one-third to
be done later. We also decided to keep the precast factory on
the site intact, saving demolition time (and cost), and let it
serve as an exhibition of the Habitat construction process.

Habitat was always about more than its purely visual
appeal or the quality of life for residents. It was also about
revolutionizing the way housing could be built and assembled.
Our ambition was to show that you could cast residential
modules in a factory operating on the site. Each module
would be a complete three-dimensional unit, with all the
constituent parts in place: doors and windows, plumbing
and wiring, bathroom and kitchen. A crane would then lift
the module into its allotted space in the larger structure,
and then all the necessary connections would be made to
utilities. I vividly remember the day
the first module was moved into place.
In a special ceremony, Nina was invited
to break a bottle of champagne on
the concrete surface, as if she were
launching a ship.

We worked with industry to
develop new products and components. For instance, we wanted to
install a one-piece prefabricated bathroom in the modules. Buckminster
Fuller had designed one made of metal
for his famous Dymaxion house in the
1930s, but such a thing had never been
produced commercially. Gel-coated
fiberglass—at the time, an up-and-

A central idea:
prefabricated
units were
constructed in
an on-site factory,
then hoisted
into place.

coming material being used for boat construction—seemed
to be the right choice. We found a company in Toronto, Reff
Plastics, that was making fiberglass furniture. Excited about
Habitat, Reff built the mold and the mock-up for the bathrooms. Similarly, Frigidaire designed a modular kitchen. The
company also designed, at our request, a stacked washer and

April 7, 1966:
Nina officially
inaugurates
the first unit of
Habitat '67.

dryer to fit into our bathroom. That saved a lot of space and was very convenient to use. Stacked washer-dryers are commonplace today but were unknown in the mid-1960s.

The streets and walkways of Habitat can be found at levels two, six, and ten, potentially exposing them to the elements, which include heavy snow and a bracing Canadian wind off the Saint Lawrence. There are places in Montreal where closely clustered buildings create wind tunnels so strong that crossing the street sometimes becomes impossible. We studied the wind conditions carefully, assessing how air would move around a complicated and porous structure like Habitat. Advising us was the head of McGill's aerodynamics department. He placed a scale model of Habitat into wind and smoke tunnels for testing. We concluded that the streets needed windshields along their length, to provide protection, and that a curved shape would be best to deflect the wind. We opted for yet another relatively new material: acrylic, which was both robust and transparent. Rohm and Haas signed on to provide what we needed.

Threats to the integrity of the project never quite disappeared. I think of one of them today as the *Chatelaine* affair, named for the Canadian women's magazine. *Chatelaine* had offered to furnish the interiors of the Habitat units in a variety of styles, an offer I vociferously opposed. One day, I arrived at the site to find an eight-foot papier-mâché sphere with funny holes in it, pink and lime in color. I asked what it was and was told that *Chatelaine* had just delivered it. It was intended to be a sculpture for the playground. I called Ed Churchill to get it removed, but he was away. So, Nina and I went down to the site that night. The sculpture sat in the moonlit plaza. The noise we made trying to drag it away attracted the security guards. Soon there were sirens, and then the head of security for all of Expo arrived. We didn't get locked up, but two further weeks of pleading were needed to persuade Expo to instruct *Chatelaine* to take the sculpture somewhere else.

An aerodynamic
test of Habitat's
transparent
walkway shields,
which deflect
winds coming off
the river.

As we got closer to the opening of Expo, public and professional attention began to mount. During the planning process I had become friends with Arthur Erickson, at the time the biggest name in Canadian architecture. He was designing Canada's Expo pavilion and several others. He invited me to a reception at his house in Vancouver to talk

about Habitat. One especially memorable moment was a visit
to the Habitat site by the architects I. M. Pei, Paul Rudolph,
and Philip Johnson, who toured the project together. Johnson,
not surprisingly, came across as the impresario—elegant and
erudite. This was some years before he adopted his trademark
eyeglasses with the round lenses and heavy black frames.
His interests were primarily visual. I remember him saying,
"You have outdone Piranesi! You have outdone Piranesi!" I'm
sure I was not the first to receive the inflated currency of
his praise. Paul Rudolph spoke softly, almost in a stutter. He
was an architect's architect and drew like an angel. He also,
I could see, was taking it all in, understanding the potential
in terms of livability, of taking a conventional residential
building and breaking it up into multiple components, and
in later years he would embrace and develop the ideas of
Habitat in some of his projects. Pei
was ever the aristocrat, appreciative
but also somewhat aloof.

During construction we also had
several visits from Buckminster Fuller.
He was responsible for the massive
geodesic dome that formed the centerpiece of the U.S. pavilion at Expo.
By then Bucky, as everyone knew him,
was a legend. He had pioneered light-

Receiving a charm
bracelet from
Buckminster
Fuller, designer of
Expo 67's geodesic
dome, during a
visit to my office
as Habitat was
being built.

weight dome construction, devised the Dymaxion house,
and conceived revolutionary new forms of mapping. He was
famous for riveting lectures on a Renaissance range of subjects that could go on for hours—I attended one at McGill.
The first time he came to see Habitat, still in model form,
was at the behest of Expo's management—they wanted to
be sure he gave the project his blessing. He did, and as he
prepared to leave after that first meeting, he took out of his
pocket a little bracelet with charms on it and gave it to me—a
comradely gesture that I have never forgotten.

Nina and I moved into apartment 904 in Habitat '67
with our children, Taal and Oren, just before the opening.
We celebrated with a party that spilled out over several of the
Habitat terraces. Music from a Caribbean steel-drum band
carried across the harbor, Montreal's skyline gleaming in the
background. Our family lived there throughout the world's
fair. It was like living in a carnival. Because one third of the

complex was open for public visits, visitors often strayed, walking into our apartment in error. One might hear a commotion and look out the window to see Charles de Gaulle, arriving to meet Expo's commissioner general, whose official residence was apartment 1011. The press coverage of Habitat '67 was, to me, astonishing: it seemed to encompass every newspaper, magazine, and TV program in North America (and beyond), culminating with a cover story in *Newsweek*. Invitations to visit and lecture poured in from around the world: India, Britain, Latin America, Israel, and elsewhere.

As is often the case, the critical reception for an innovative architectural project such as this one was mixed. Ada Louise Huxtable, the architecture critic for the *New York Times*, described Habitat as the promise of things to come. She opened her story by describing a mailman walking along one of Habitat's streets and delivering mail—and then noted that the street is a "skystreet," ten stories in the air. She went on: "Just about every housing and building rule, precedent, practice, custom, and convention is broken by Habitat." The project, she wrote, "is a significant and stunning exercise in experimental housing that is also the most important construction at Expo, where architectural excellence abounds." Some of the more academic critics, like Reyner Banham, were skeptical, arguing that it was not a reproducible model, and was ill-suited to Montreal weather. Banham called Habitat '67 "a fifth-year student's thesis that somehow managed to get built." This wasn't entirely accurate—it was a *sixth*-year student's thesis.

Louis Kahn's response to the journalist writing the cover story for *Newsweek* was curiously standoffish. Maybe that was understandable: I not only had hired away his engineer, August Komendant, but had won international recognition at a young age and with my first project. Kahn struggled for decades before winning the appreciation he truly deserved. Whatever the explanation, Kahn and I would meet cordially in the future in various places. My debt to him is considerable.

When Expo closed, at the end of 1967, permanent residents began moving into Habitat and putting down roots. Among the residents was the Gopnik family, whose children include the writers Adam and Blake. A few years ago, when Habitat turned fifty, Blake Gopnik wrote a retrospective essay in the *New York Times*:

When I was 5, I saw our new home as extending endlessly in every direction. It soared above a vast, churning river on one side and a hectic port on the other. It was exactly the building my kindergarten self would have built, stacking blocks to the verge of collapse. Its pleasures left me breathless the first day I woke up there and most mornings after. . . .

Built on a spit of land separating the rapids of the great St. Lawrence River from the working waters of Montreal's port, Habitat's 158 apartments fill 354 cast-concrete boxes, piled 11 stories high in a madcap mess of cantilevers and bridges and perilous open spaces—like (guess what) a stack of children's blocks. For sheer sensory excitement, Habitat could not and cannot be matched. Every minute in the building felt unlike the next, as space, light, air and sound danced around you. My parents built a jungle-gym on one of our terraces, but the building was the best climbing frame of all.

Cover story, 1971.
The publicity,
worldwide,
continued for
years.

Prayers at the Western Wall in Jerusalem, 1920. Back then, men and women were not separated.

CHAPTER FOUR

Old City, New City

Midway through the world's fair, something happened that would have a profound impact, in time, on my personal and professional life: the Six-Day War. In June 1967, Israel faced the combined armies of Egypt, Jordan, and Syria, and defeated them all in the space of a week. Experiencing this in Montreal was terrifying and emotional. My friends from the Reali School, in Haifa, were all at the front. One of my close childhood friends, Mordechai Friedman, was killed in the battle for Jerusalem. I had not been to Israel for fourteen years but had remained in close contact with my peers. To be absent at a time of crisis filled me with a sense both of guilt and anxiety, even if there was nothing I could conceivably have done. I reported to the Israeli consulate, offering myself as a volunteer, but the war was over quickly.

Israel, from that moment, was again very much on my mind. And in December of 1967, I was able to return for the first time since my family had left. In the euphoric aftermath of the Six-Day War, when the troubling long-term consequences of that conflict were perceived by only a prophetic few, an international conference of architects and engineers was being held in Tel Aviv. Because of my Israeli background and the attention that Habitat had received, I was invited to be a keynote speaker and eagerly accepted. Nina, Taal, Oren, and I arrived to find a country vastly different from the one I had known.

For me, like all Israelis—and all Palestinians—the Arab-Israeli conflict has been an integral part of my life. We each view it through our specific circumstances and perspectives. Not having experienced combat, I have never felt myself in the role of a warrior in the conflict, nor as an advocate of one side or the other. I have always believed that my personal history enabled me to understand the issues and concerns on both sides. That said, I have fully and always identified as an Israeli.

Haifa, my childhood home, was one of the more mixed communities in Palestine under the British Mandate, as it still is today. The scenes surrounding its "liberation"—or "conquest," depending on your point of view—by the Haganah, the precursor of the Israel Defense Forces, in 1948, remain vivid in my memory. The War of Independence felt like a war of survival; almost every family we knew suffered causalities. The siege of Jerusalem, the invasion by multiple armies: all this impressed on a young person how vulnerable we were. That sense of vulnerability informs the Israeli psyche to this day. It is multiplied by the experience of the Holocaust and by Arab rhetoric. As Israel became stronger and more established, and militarily more powerful, one might have expected the psychological dynamics to change. Sadly, they have not.

With the start of the Six-Day War, a combination of emotions washed over me: my fourteen years of longing for Israel, to start with—now intensified—but also guilt at having been studying architecture in a safely distant country, and guilt at never having served in the army, while my friends had put their lives on the line. In later years, when I returned to work in Israel for a significant period of time, I would indeed join the army and serve in the reserves. I never had to fight. In Israel, when people allude in passing to what they did in this or that war—which happens frequently—I feel like an outsider.

Arriving in Tel Aviv for the international conference, I immediately felt a deep sense of homecoming: the language, the climate, the friends, the food. Jerusalem was united once again, and you could go everywhere in the city, as I had been able to do when I was young. Of course, returning to a place that was once familiar can be as disorienting as visiting an unknown place for the first time—perhaps even more disorienting.

I remember one palpable reaction: that some of the beauty of the country had been spoiled because ugly, five-story housing complexes—we called them *shikunim*—had been built very rapidly in the late 1950s to accommodate an influx of Jewish immigrants to Israel, mostly from Arab countries. They seemed to have sprouted everywhere, and to my mind offered a lesson in how not to pursue the very ends I had been seeking to achieve with Habitat. Returning to Israel as an adult, I was struck by the demographic alteration. I had been the only Sephardi in a class of Ashkenazim at the Reali School. But the new immigrants were predominantly Sephardim, and their presence was notable. Soon after arrival, as I began to talk and interact with other Israelis, I also began to get a sense, as I did not have when I was younger, of how small the country really was, particularly back then, in 1967. I don't mean just geographically, though Israel is certainly small in terms of size, but socially: the circle of professional, business, military, and political leaders was tightly interwoven. On the whole, though, the country seemed less glamorous and more provincial than I remembered. I had to get out of the cities and into the agricultural regions before I could encounter, once again, the extraordinary beauty I had remembered.

At the architecture conference in Tel Aviv I met, for the first time, my peers in Israeli architecture—people such as Yaakov Rechter, Ram Karmi, Ada Karmi-Melamede, Avraham Yaski, and Amnon Niv. Many would become good friends. Prominent international figures were also in attendance: Richard Meier, Buckminster Fuller, Philip Johnson, and my mentor Louis Kahn. It was a moment of optimism—a word not often associated with this part of the world. On its last day, the conference moved from Tel Aviv to Jerusalem, and on this occasion, I met a man who would become an important part of both my life and my career: Jerusalem's mayor, Teddy Kollek.

There was something seductive and disarming about Kollek. His face was rather round, the cheeks rosy and glowing, always a hint of a smile. He had a thick head of light brown hair. He was a little heavyset, but agile. He would focus his gaze on you, fully concentrating. We hit it off right away, and his interest and collaboration gave me an architectural foothold in Israel, one that at first did not always yield results, and in fact led to many bitter disappointments, but that has proved to be profoundly meaningful.

Kollek would never have struck anyone as a typical Israeli. He was originally from Vienna and had compiled an illustrious record during the Second World War. On behalf of the Jewish Agency, he worked with British and American intelligence. He helped facilitate immigration from Europe to Israel. He bought arms and ammunition for the Haganah. There was an unmistakable mystique about him. He was a member of a kibbutz, but he smoked fine Havana cigars and had the sophistication of a European intellectual. That wasn't the way most Israeli leaders of those early years came across. There was an old joke that David Ben-Gurion's wife, Paula, asked by a friend if they'd be attending a performance of Beethoven's Fifth, had replied, "Oh, we've already heard that one."

In the early days, Kollek had been the chief of staff to Ben-Gurion and, as a result, was at the center of the decision-making for the entire country. Kollek created the Israel Museum, which today occupies a hill not far from the Knesset; the museum is one of the world's finest repositories of art and archaeology. Kollek was a legendary fund-raiser and seemed to know everybody, not just in the Jewish world but in the philanthropic world more generally, and on several continents. Then, in 1965, he ran for mayor of Jerusalem and won. For all its emotional resonance, Jerusalem was not a major metropolis. It was provincial, with half the city, including all of the Old City, under the control of Jordan. Kollek was drawn to Jerusalem not for reasons of political power but out of interest and love—he was determined, as he once said of the city, "to take care of it and show better care than anyone else ever has."

Jerusalem's mayor, Teddy Kollek, in the 1970s.

And then, as a consequence of the war, the city was unified overnight. The Old City again lay at its heart. To the west of the Old City's walls ran a wide, north–south gash of what had once been no-man's-land—a scarred terrain of concrete walls and barbed wire that had been the scene of fierce fighting. Kollek had to figure out what to do with the united city that he now had on his hands, a third of its populace Arab. He created something called the Jerusalem Foundation and raised countless millions to build schools, parks, and other public facilities. There was a boom in construction; people used to say he was the greatest builder of Jerusalem since Herod. Hanging over everything was the larger question: What was the right way to develop this ancient and holy city?

From the moment of our meeting, Kollek sought to draw me into his plans, which meant involving the Montreal office as well. Many things happened at once. The minister of housing and construction, Mordechai Bentov—a tanned kibbutznik in an open-collared white shirt who looked like a fantasy version of an Israeli founding father—wanted us to bring the Habitat concept, along with the technology that underlay it, to Israel. He assigned us a site and a contract. There was an immense amount of work to be done in the Old City.

The Jewish Quarter had been heavily damaged during the 1948 War of Independence—synagogues and seminaries destroyed, for instance—and its Jewish residents had been expelled. Arab squatters had lived among the ruins for two decades. Now, a corporation was set up to restore and repopulate the Jewish Quarter, and I and my firm were given portions of the quarter to rebuild. On top of that, we were commissioned to rebuild a Sephardic yeshiva that faced the Western Wall, the last remnant of the temple that the Romans had reduced to ruin in 70 CE. The yeshiva itself, which opened in 1923, had been destroyed during the War of Independence, blown up by the Arab Legion. Relatives of mine, the rabbis who headed the yeshiva, were involved in the rebuilding effort. We were soon at work on that as well. And then, finally, in 1972, Teddy Kollek asked that we take on the planning for the reconstruction of Mamilla—that twenty-five-acre tract of former no-man's-land outside the city wall, between what had been the Israeli and Jordanian sectors of the city.

Israel's housing and construction minister, Mordechai Bentov, 1966.

So, Israel quickly became a place to focus my energy and attention. With a foot replanted there, I knew that it would remain planted. In 1971, I opened up an office in Israel as a companion to the office in Canada, hiring several young Israeli architects to staff it. Within a year or two, I was commuting regularly between the two offices, sketching furiously in my notebooks during the long international flights.

* * *

The Montreal office was busy with other projects too. After Habitat '67 was completed, Nina and I found ourselves, for the first time, with an income beyond basic sustenance. I allowed myself one indulgence. Succumbing to a love for cars—which I trace back to my father's Studebaker—I bought a 1968 Citroën DS21 convertible, a limited edition issued by the French designer Chapron, and one of the most beautiful cars ever made. I believe only a hundred or so were manufactured. I've kept the car to this day, maintained and restored over the years. I still drive it in the summer, though it now qualifies as an antique. I'm not sure there was a connection, but Leonard Cohen produced one of his videos, for the song "In My Secret Life," while singing within Habitat '67, a Citroën parked in the foreground.

For all the activity on the Montreal side of the firm, the results proved disappointing in the short term. Many promising pathways turned out to be cul-de-sacs. The experience would prove to be a real-world lesson in how difficult new ideas can be to implement, even when "new ideas" are what people say they want.

I had taken real-estate developers and heads of state alike through the Habitat '67 complex. Feelers about the possibility of Habitat-inspired projects began coming in, sometimes even before the world's fair officially opened. The first of these was particularly intriguing—a letter from the student committee of San Francisco State College, which was about to build a new student union in the heart of the campus. This was at the height of the so-called counterculture, centered in Berkeley and San Francisco. The student union was being built with student funds, raised from membership dues accumulated over the years, so the students would rightly have responsibility for choosing the architect. The design, however, would require the approval of the state college board, because the student union would sit on state land. The governor of California at the time was Ronald Reagan.

My 1968 Citroën DS21 convertible—still running.

I traveled to San Francisco for an interview. I was almost the same age as some of the students, and the chemistry was good. Our firm was chosen, and the design that evolved was radical: a structure assembled from precast forms that were bent, stacked, and clustered to form a hill-like building you could climb over, walk under, or enter directly. Many of the inclined surfaces on the exterior were given over to stairways, and much of the rest of the exterior was planted with grass. The student union we envisaged was a verdant, crystalline structure—referred to in shorthand as the "grass and glass" building. I traveled with the student committee to Los Angeles to present the design to the state college board. It was an unforgettable scene: arranged on one side, a group of students looking like refugees from Woodstock, along with one Canadian Israeli with a mustache; arranged on the other, a phalanx of white men in suits. Maybe predictably, the design was rejected by the board, a rejection

Model of the "grass and glass" student center proposed for San Francisco State College, 1968.

that became one of the many issues precipitating student riots on campus soon after.

I got caught up in the protests by accident. As I was commuting between Montreal and San Francisco, I had fallen into the routine of staying at the Berkeley home of the architect and design theorist Christopher Alexander, who had become a close friend. Chris and I shared basic values and a taste for particular burgundies, though we approached architecture quite differently. During one of these visits, the demonstrations over People's Park in Berkeley reached a crescendo. A curfew was imposed. After dinner, Chris and I got into his Porsche convertible and set out to explore. I can't reconstruct what we could possibly have thought we were doing. As we drove toward People's Park, we were suddenly intercepted by two police cars. We were pulled from the car, handcuffed, and taken to a large cell in Berkeley's jail, which at that moment held some fifty other people, many of them students, some of them regulars. An exposed toilet occupied the center. I kept asking to see the Canadian consul, as if that was likely; they must have thought I was nuts. In the morning we were arraigned and then released, with no charges filed.

Amid all the student unrest, San Francisco State's president was replaced by a hard-line professor named S. I. Hayakawa, a linguistics scholar who wore a jaunty tam-o'-shanter and went on to become a conservative hero and a U.S. senator.

Meanwhile, in 1967, I had a call from Manhattan. Carol Haussamen, a wealthy real-estate owner in New York who was very close to Mayor John V. Lindsay, was leading a group of urban designers to rethink development in the city. Lindsay had visited Habitat '67, and Haussamen was encouraging him to build a version of Habitat in New York. Two city-owned sites were offered as potential locations—the old asphalt plant along the FDR Drive, north of Gracie Mansion, on the East River, and the piers below South Street Seaport, also on the East River, in lower Manhattan. The second of these was the more interesting and promising site. It was on the river side of the FDR Drive, with great views toward Brooklyn. Because the desired density demanded a structure that would be thirty or forty stories high, the modules could not be load-bearing—stacked directly one on top of another—as they had been in Montreal. In the aggregate, they would simply be too heavy, and the load therefore too great. As an alternative to inserting the modules into an orthogonal steel support frame, I turned to the idea of suspending the structure on cables. The affinity with the nearby Brooklyn Bridge, one of the earliest suspension structures ever built, was uncanny. The engineering challenges would be significant, though not insuperable—but they were enough to scare off potential contractors. In the end, the only thing that ended up being suspended was the project.

Two other Habitat-inspired enterprises were initiated by the U.S. Department of Housing and Urban Development. HUD today is mainly a landlord and a source of subsidized loans. There was a time, in its early years, when it was a source of innovation, even under a Republican president. In 1969, George Romney, who had headed the American Motors Corporation and been governor of Michigan, was appointed

Property receipt, Berkeley, California, 1968. Christopher Alexander and I were released after a night in jail during student protests.

PRISONER'S PROPERTY RECEIPT

HUD secretary by President Richard Nixon. Romney's mission, indeed, passion, was to introduce new, industrialized systems to housing; after all, he was an assembly-line expert.

Romney set up a program called Operation Breakthrough to encourage experimental ventures—including Habitat projects in the Fort Lincoln neighborhood of Washington, D.C., and another on the outskirts of San Juan, Puerto Rico. I knew the cost of construction would be critical; to achieve a lower cost we had to come up with lighter modules. We joined forces with a structural engineer I had come to know, Ed Rice, who was a partner at T. Y. Lin, the global engineering firm. Rice had developed a special, expansive concrete that could "prestress" the modules, making them strong enough to allow three-inch walls to do the same job as Habitat's five-inch walls. Each module would weigh thirty tons instead of seventy. In Puerto Rico, a plant was set up on-site, and modules started coming off the production line.

Model for a proposed Habitat New York in lower Manhattan, 1968.

But there was trouble at HUD. Long-standing disagreements between Romney and Nixon—including over the desegregation of public housing—eventually led Romney to resign. Operation Breakthrough was abandoned, and so were the Habitat projects. Today, some of the modules can still be seen on an overgrown hill above San Juan, looking like an archaeological site in the jungle. Ironically, when I appeared on the cover of *Newsweek* in 1971, the model that took up most of the background was that of Habitat Puerto Rico.

Another victim of federal cutbacks was an ambitious project initiated by the Baltimore Development Authority, led by Robert Embry. He had read my book *Beyond Habitat*, a chronicle of the Expo project and a personal statement of ideas and beliefs about housing that I had published in 1970. Embry invited us to lead the planning and design effort for a new community in Baltimore called Coldspring New Town. The development, in a wooded area centered around a disused quarry, would have nearly four thousand units. The Rouse Company—a developer that was then building the city of Columbia, Maryland, and would later develop Faneuil Hall in Boston and the Inner Harbor in Baltimore—was brought

Design for Habitat
Puerto Rico,
1971, backed
by the federal
government
as part of an
innovative
program—then
abandoned.

in as a consultant. We spent two years on the master plan
and architectural design. Community meetings, involving
representatives from affluent white neighborhoods to the
north and lower-income Black neighborhoods to the west,
were often intense. About a third of the units were desig-
nated as affordable housing. Eventually the plan won approval,
and construction began. But Washington pulled the plug on
funding. Only about two hundred units were ever built. It's
a desirable place to live but nothing like the community it
could have been.

These early years were enhanced by an invitation I
received early in 1971 from Yale University to become the
Davenport Visiting Professor at the School of Architecture—a
prestigious chair, which has been held by James Stirling and
Richard Rogers, that affords the opportunity to teach for
limited periods of time. The class of sixteen students was
impressive, and I thought hard about what I might give them
as the subject for their studio—their culminating project.

Traditionally, a design problem is assigned, and each student works to come up with a scheme in response, getting
critical input from the faculty. Asking the students to design
a housing complex, so as to engage with the spirit of Habitat, did not strike me as sufficiently challenging. I wanted
something new and unfamiliar. As luck would have it, the
international competition for what would become known as
the Centre Pompidou—at first called the Plateau Beaubourg—
in Paris had just been announced; it was among the most
important international competitions since the one for the
Sydney Opera House, in Australia. I initially decided to take

this as our subject. Then I had another thought: rather than do this as a conventional studio, with each student developing his or her own design, we could all pretend we were partners in a firm, developing a design that we would actually submit. This meant that I, too, would have to roll up my sleeves and be part of the team.

By semester's end, we had an exciting proposal, but it was not yet ready for presentation. There was still a month to go before the submission deadline. The entire class moved up to Montreal and squeezed into our temporary house in Westmount—a rambling mansion with a huge attic that we were occupying temporarily. That we even had such a place to offer was perhaps the one silver lining in what was otherwise a traumatic event. In 1967, Nina, Taal, Oren, and I had left our unit in Habitat and moved into a white-painted, brick-clad Victorian house on the slopes of West-mount. On a bitterly cold and snowy night in 1972, we had been having din-ner with friends, while a fire burned in the fireplace. Unbeknownst to us, a crack had developed in the fireplace bricks. The fire spread quietly in the walls. At three a.m., Oren came into our bedroom coughing. We were out in seconds, the house bursting into

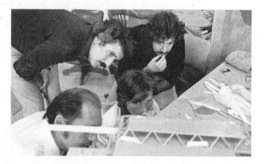

flames behind us as fresh air rushed in through the door we opened. Everyone was safe, but our possessions were gone. We moved into the Westmount mansion with the attic before returning to live in Habitat.

Working with Yale students on a submission for the Beaubourg competition, 1971.

My students and I, living and working together, com-pleted the design for Beaubourg. It was unlike anything one would expect a museum in Paris to be. If approached from one of the surrounding streets, the building would resemble a hill, planted over with greenery and offering a series of public stairs that ascended toward the top. Those arriving from other directions would come upon two large, dramat-ically cantilevered structures, one crossing underneath the other. The two structures spanned the site and opened into the underside of the building, where the ground descended into terraced galleries (and to the Metro). In the end, the winning design was the one submitted by Renzo Piano and Richard Rogers, and their Centre Pompidou is today an iconic presence in the old Les Halles neighborhood. Our own design was awarded one of several second prizes. We didn't make it, but we got close.

The series of projects that came into the Montreal office in the first few years after Habitat '67 had been extraordinary in scope. I was reinventing Habitat over and over, adapting it to each particular context. The San Francisco student union and the Centre Pompidou schemes are both recognized today as groundbreaking, although neither was realized. But the number of false starts and dead ends was discouraging. Almost nothing got built. After Expo, I had come to believe that the success of the first Habitat would bring many more Habitats to life. What was becoming clear to me, experiencing one disappointment after another, was that Expo had been a special, one-of-a-kind event. At least in North America, innovative projects in the "normal" world would for a time prove maddeningly elusive.

* * *

Jerusalem was a different story. The city lies some twenty-five hundred feet above sea level, and the road to Jerusalem from the coastal plain rises steeply and unforgettably into the Judean Mountains. A first glimpse of the Old City, set on a plateau surrounded by hills, is a sight no one forgets. The impact is not merely visual or sensory. As a city held sacred by the three Abrahamic religions, Jerusalem tugs at the soul, regardless of one's religious beliefs or lack of belief, and regardless of what one thinks of the word "soul." For me, by far the most beautiful moments in Jerusalem are those when the golden-hued city glows at dusk, and the sounds of church bells ringing, muezzins chanting, and Jews praying

A Beaubourg cross-section. Richard Rogers and Renzo Piano got the honors, but our students won a second prize.

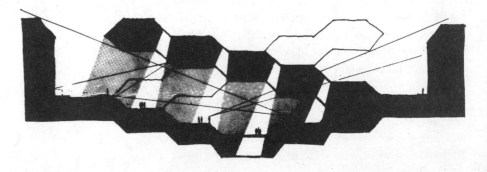

at the Western Wall all blend together. I recall Saul Bellow's comment: "There are many Israelis who do not believe, but there are few who have no religious life."

I went back and forth between Israel and Canada, and because many of the government officials I initially dealt with were in Tel Aviv, that's mainly where I was at first based. Nina and the children remained rooted in Montreal but in the early 1970s would spend the summers in Israel. Tel Aviv was not yet the vibrant, cosmopolitan city that it would become, with an identity so distinct that it is sometimes referred to as the State of Tel Aviv. But it was lively and commercial and the secular heart of the country. Whether in Tel Aviv or Jerusalem, I began to develop friendships not only with leading local architects but also with journalists and people in the arts, and with writers such as A. B. Yehoshua and Yoram Kaniuk. Zubin Mehta, whom I had met as the young conductor of the Montreal Symphony Orchestra, was now beginning his activity with the Israel Philharmonic.

Each of these relationships opened windows into different facets of Israeli life, including its business community, the government, and even the military. Yoram Kaniuk, for instance, had written a celebrated novel of the Holocaust, translated into English as *Adam Resurrected*. (The Hebrew title, translated literally, is *Adam, Son of Dog*.) One of his close friends was the commander of the armored forces, Israel Tal. Among the most respected generals in the military, Tal was credited with building up Israel's formidable armored capability.

I remember a night when the discussion turned to my book *Beyond Habitat*, in which, among other things, I had argued that seeking beauty as a design objective unto itself was doomed to failure. Rather, seeking a design that was perfectly fitted to purpose would lead inevitably to beauty. Tal—or Talik, as he was called—was skeptical, but eventually he laid down a challenge that led to a design effort that was not "architectural" in the proper sense of that term. Tal revealed, "I'm in the process of designing a new tank. I want it to be the most beautiful tank in the world. I'm going to involve you. Let's test your theory." The tank was to be called the Merkava—Hebrew for "chariot." Its design and production were closely held secrets. But the next thing I knew, we were in a Jaffa warehouse studying a plywood mock-up of the tank.

I immediately took in the most radical aspect of its design. All tanks, until then, had their engines in the rear, which meant that one could get inside the tank only through the turret. Talik had placed the tank's engine in the front, allowing the engine to provide a protective layer for the crew but also making it possible to place a door in the rear, allowing troops and crews to exit and enter the tank under cover. The mock-up was covered with wartlike protrusions, as if it were suffering from bubonic plague; the warts accommodated vents for air intake and exhaust as well as attachments with sensors of various kinds. I hadn't given much thought to what would make a beautiful tank, but I reflected on my Citroën. Its beauty had to do with its streamlined look. The headlights, taillights, and many other elements somehow found a way into its aerodynamic envelope.

I set out to understand the purpose of each of the Merkava's protrusions as well as the parameters for the overall dimensions and shape of the tank. It was clear that the turret and front engine could accommodate a natural aerodynamic form. The back would be more like the aft of a ship. My focus over the next several visits was to figure out how to incorporate the protrusions into a streamlined envelope. Functionally, this was necessary: the protrusions represented a hazard. Shells, which might bounce off a smooth surface, could snag and detonate. By the time we were done, there wasn't a single wart on the exterior. And the functional necessity had an aesthetic payoff: the tank possessed the sleekness of a bullet.

In the end, the person who had really proved my theory was Talik himself. He had gone back to first principles, reflecting on his years of battle experience. He had come to understand the advantages of being able to enter the tank from behind. Everything flowed from that insight. It was a perfect example of design fitted to purpose.

* * *

Over time, the center of gravity of my work began to shift to Jerusalem. Instead of modern Tel Aviv hotels, I began staying at the historic American Colony Hotel, north of the Damascus Gate, a place that would prove to have considerable impact on my thinking about Jerusalem because of its architecture, its environment, and its indescribably evocative character. For my Jerusalem office, I hired as a personal assistant a woman who had been Teddy Kollek's right arm for many years, Yehudit Yaacovi, who was herself very well connected in the city. Within a few years, I built up an Israeli staff of about fifteen people. Certain interests and skills came together for the first time in a seamless, holistic way—an interest in planning large communities, an interest in designing individual buildings, an interest in social dynamics, an interest in context, and an interest in that ineffable thing you might call the spirit of a place.

The Merkava tank, 1979, developed by General Israel Tal.

I have found over the years that people are often confused about what differentiates professions such as architect, urban designer, and city planner. Even in academia there is no clarity about whether these are different disciplines or in fact are all part of some larger, single enterprise. I have come to think of myself as an architect and an urbanist, or maybe as an architect and an urban designer. Whatever the terminology, the elements are intimately related but also distinct, with different processes and different ways of addressing the creation of our environment.

An architect designing a building has a finite assignment. A building is, so to speak, a singular organism. One can think of buildings in terms of typologies: a house, a residential complex, a hospital, a museum, an airport. Each type has particular issues it must resolve. Knowledge of what is called "the program" is fundamental to the ability to design a structure that addresses spatial needs while also addressing broader technical and contextual issues. Thinking about the program is partly quantitative: How much space must be devoted to which kinds of activities, and how do the spaces relate to one another? Which spaces are publicly accessible, and which ones need to be secure? What are the special performance requirements? Labs, for instance, need floors that don't absorb—for purposes of hygiene and safety—as well as extraordinary ventilation. But qualitative thinking is also centrally involved at the program stage. What materials and methods should be deployed? Is it important for hallways to be silent? Should windows look out onto courtyards? Where are the zones of interaction? What color palette might be appropriate? And then, an even larger consideration: How does a structure relate to its surroundings? In that respect, while not acting as an urban designer, an architect must be cognizant of the impact of a building on what is around it, and on the public realm that abuts it.

Nowhere does the idea that a building should relate to its context come more urgently to life than in Jerusalem. With such a rich historical, cultural, and architectural heritage, the challenge is to achieve contemporary, modern design that is harmonious with its setting. Responding to that challenge transformed me as an architect.

An urban designer needs to think differently from an architect. When in 1972 I first came to develop a master plan for the twenty-five-acre Mamilla district, I realized that my task was not a matter of designing a collection of buildings. Rather, I had to switch hats and become something else. When a district is designed, there must be an overall program, just as there is for a single building. One needs to think about area requirements for each building type—apartments, offices, hotels, shops. Space must also be allocated for roads, walkways, and infrastructure. Designing an individual building is like placing a building block into an existing urban context. In contrast, the master-planning process consists of creating the urban context itself: arranging an assortment of building blocks that can be manipulated to create a satisfying public realm.

Put another way, the urban designer's prime responsibility is to design the spaces *between* the buildings rather than the buildings themselves. The character of those spaces must serve the community in the manner intended. Needless to say, there is an intimate interdependence between architecture and urban design.

Jerusalem, when I arrived, was badly in need of both specific individual buildings, to replace those that had been destroyed and to accommodate the sudden and extraordinary growth, and also new kinds of urban design—to rehabilitate ravaged neighborhoods and fill empty stretches of desolation. It was a city with a distinctive vernacular architecture that could help define a path forward: low buildings, a gnarled streetscape, an almost epidermal relationship with the terrain, and everywhere, golden limestone. It was

also a city that needed preservation and protection: a city
that did not want to be Tel Aviv, with its skyscrapers and its
brash modernity.

I admired Teddy Kollek's appreciation for what it takes
to make a city great. He was interested in parks, in culture, in
beauty, but he was equally interested in education and other
social issues, in history, in archaeology. In exchange for the
few cigars I'd occasionally buy for him when passing through
Zürich Airport, Kollek sometimes gave me small artifacts
that he owned—an ancient bronze figurine, say, or a pottery
lamp. I cherish my little archaeological collection. He was
indefatigable and needed three secretarial assistants: one
who started at the city office at dawn, a second who came
at midday, and a third who took over in the evenings when
he worked from a private office at the Israel Museum. Like
many Europeans, Kollek was a prisoner of preconceptions
and prejudices about Palestinians, and yet, unlike many who
followed, he reached out to the Arab population, bringing
social and municipal services. Among other things, through
the Jerusalem Foundation he sought outside donors to spon-
sor projects specifically for Palestinians. One of them—the
Youth Wing our firm designed for the Rockefeller Museum,
in East Jerusalem, in the heart of the Palestinian sector of
the city—was funded by William S. Paley, the onetime chief
executive of CBS, the American radio and television network.
It would become a major arts center for mixed groups of
Arab and Jewish children. (Alas, the fears instilled by two
Palestinian uprisings—the intifada that began in 1987, and
a second that began in 2000—curtailed Jewish enrollment,
and the center now serves Arab children exclusively.)

As I started commuting to Jerusalem monthly, I got into
the habit of getting up early and visiting Kollek in his office
in the old Mandatory municipal building at six a.m. He was
an early riser, and I knew these would be quiet moments
before the staff arrived. The first few years after 1967 were
turbulent in terms of city planning in Jerusalem. Prior to
unification, at the behest of the municipality and the national
government, a master plan had been prepared for the city
with the help of U.S.-based transportation consultants. The
plan could not have anticipated unification, but essentially it
called for a system of expressways in the old urban-renewal
style of the 1950s and '60s—the kind of approach that had

ripped the heart out of Seattle, Boston, San Francisco, and other places. This was the approach upon which Jane Jacobs had heaped scorn in her classic book *The Death and Life of Great American Cities*.

Kollek realized that he had a problem on his hands. He had the brilliant idea of convening a Jerusalem committee, an assembly of the leading architects and planners from around the world, to act as an advisory body. I was a member, along with figures of far more prominence, such as Buckminster Fuller and Louis Kahn. The committee helped bring about the demise of the old master plan and helped Kollek introduce sounder polices. As the *Jerusalem Post* observed at the time, "The foreign critics were not wielding a scalpel on the master plan, but a guillotine."

Luckily for Jerusalem, Teddy Kollek had no aspirations for any office besides mayor. But he had the stature of a states-man, someone who can lead with an inner, self-motivating voice, not one who follows the polls. I've worked with many politicians, from mayors to prime ministers—and once, an empress—and it quickly becomes clear which ones are driven by conviction rather than opportunism. Only the former produce transformative results.

* * *

Many of my early individual efforts in Jerusalem did not come to fruition—but over time, others did: Hebrew Union College, the Jewish Quarter, Yad Vashem, and the National Campus for the Archaeology of Israel. The ambi-tious Mamilla project, with its sensitive location and vast scale, required decades of persistent work before the dream was finally realized, as it now has been. Every step demanded the exercise of politics and diplomacy at many levels. Often, politics and diplomacy proved unavailing. Looking back, I realize that Jerusalem was, for me, as it has been for countless others down the ages, a great teacher.

The lessons were often painful. The very first project that had been proposed to me—for a Habitat outside Jerusalem—fell victim to what might be called geopolitics. We had orig-inally chosen a location on the edge of Jerusalem but within the Green Line of Israel—that is, within the pre-1967 borders, not in territory occupied after 1967. But then the national government decided that its priority was to build a series of new neighborhoods interspersed with Arab neighborhoods— new satellite towns built in the now-expanded city limits, formally annexed to prevent any future division of the city. These new towns represented political statements, and their architectural components were rushed. Mordechai Bentov, the housing and construction minister, was retired by then. I did not wish to get involved with projects in the new neighborhoods. Habitat Jerusalem came to an end.

Geopolitics impeded another visionary effort—an attempt to build housing for Palestinians. My interest in this went back a long way. As noted, in the early 1960s, I had often spent evenings sketching ideas for a city of 150,000, intended explicitly for Palestinians made homeless after 1948 and now living as refugees in the surrounding region. My plan called for a series of pyramidal structures along a linear urban axis, and I imagined that the new city could be built near Giza, on the outskirts of Cairo—not taking into account the resistance of all Arab countries, Egypt included, to absorbing Palestinian refugees. In retrospect, I wonder at my sheer naivete.

After 1967, the Palestinian question was again on my mind. The refugee situation was now more severe. In 1970, I thought there was an opportune moment for major new efforts to house at least some of the refugees in lands now controlled by Israel. I had recently come to know Lord Victor Rothschild, the head of the London branch of the Rothschild family. He had had some business in Montreal and, knowing my work, had contacted me. Then the research coordinator for the Royal Dutch Shell Group, he had been an intelligence officer during World War II. Looking at the man in the soft, pin-striped, double-breasted suit, a slight rosiness in his cheeks, one would not have guessed that he had personally defused bombs and other explosives. Lord Victor and I realized that we shared a concern for the Palestinian refugees and decided to proceed together. The Rothschild name was legend in Israel and carried weight. Every Israeli city, it seemed, had a Rothschild Boulevard. With the help of the West Bank military command, we explored possible sites near the Palestinian cities of Nablus and Ramallah. The idea was to build a number of factories that would produce prefabricated modular units. Palestinians would be trained and employed to do the construction, and multinational companies, in turn, would be encouraged to bring industry to the new towns. Shimon Peres, who was then the Israeli minister of transport, and whose portfolio included special responsibility for the occupied territories, became an ally of the scheme. The project was presented to Levi Eshkol, Israel's prime minister at the time.

And there it was stopped dead in its tracks. As Peres reported back in a phone call, the prime minister believed that no initiative of this kind, involving billions of dollars, could be initiated before there was a full peace agreement with the neighboring Arab states. Such is the tragedy of the Arab-Israeli conflict: the antagonists insist on all or nothing, and "nothing" is the inevitable result.

One of the most exciting projects I embarked on during those first years in Israel—and eventually one of the most nerve-racking and maddening—was the rebuilding of Yeshiva Porat Yosef, the rabbinical college that had been blown up in 1948. The original structure, its cornerstone laid in 1914, had looked out over the Western Wall. Now it was largely a ruin. The head of the yeshiva, which was an ultra-Orthodox institution, was a cousin by marriage from an Aleppo family. The man who really got me involved and gave me a formal commission was Stephen Shalom, also from a Sephardic Aleppo family. He lived in New York and was a man of means with a philanthropic spirit. His father had been known as the Handkerchief King.

It was a delicate moment because, as I learned, already one design had been commissioned and rejected: a big mod-

Yeshiva Porat Yosef, in Jerusalem, as it appeared in 1923 (top), and at the moment of its destruction by Jordan's Arab Legion, 1948.

Model of the
design for a rebuilt
Yeshiva Porat
Yosef, 1970.

ernist stained-glass box designed by a New York architect.
When I first met with the rabbis, they asked me point-blank
whether I would design a modern building for them or a
traditional building. I realized the question was loaded and
spontaneously replied, "Well, if I succeed, you won't be able
to answer the question."

At first, the relationship among all the parties was everything one could ask for. I came up with a design that attempted
to reconcile the tradition of massive stone architecture with
a contemporary, lightweight structure of precast concrete
forming arches and half arches. Thick stone walls surrounded
the building, with passages, stairs, and light shafts carved
into them. A lacelike lattice enclosed many of the interior
spaces. Terraced rows of domes, some translucent and others
transparent, crowned the larger chambers, such as the synagogue. The design and the materials resonated powerfully
with the geometry of the historic city. Significantly, the large
complex was broken into many smaller-scale components, in
keeping with the surroundings. My plan won rapid approval,
and we started building.

Then things became complicated. Money issues required
bringing in new donors. Shalom, the major benefactor, insisted
that the yeshiva's curriculum must include basic secular education, including math and languages. He also wanted a gym.
The conservative rabbis would have nothing to do with such
modern ideas. This disagreement was a precursor of later
disputes that would roil the Israeli government, such as the
attempt to introduce language instruction and basic math
and science into religious schools. In this case, the rabbis

Razing of
Jerusalem's
Mughrabi Quarter
to clear space
in front of the
Western Wall,
1967.

turned to other benefactors. Disputes broke out over what
would be named after whom. The rabbis wanted to cancel
the gym. They wanted a menorah carved decoratively on
the facade. They wanted stained-glass windows. No act of
vengeance or dishonesty seemed beyond the rabbis as they
pursued whatever their objective was at the moment. I was
caught in the middle of all this. At a certain point, the rabbis
began, unilaterally, to decree changes of their own devising. I
arrived one day and discovered that the contractor had been
given instructions counter to my own.

That, for me, was the last straw. I had been trying to
keep the combatants apart or hold them together, depending on the situation. No longer. I went to court asking for
an injunction to stop changes to the design. The case went
all the way to the Supreme Court of Israel, and in the end
I won. The decisive point in my favor was a clause in the
contract that the rabbis had insisted on, stipulating that I as
the architect bore all responsibility for the performance of
the building. This had been intended by the rabbis as a way
of shifting full liability from them onto me, which indeed
it did—but the clause also backfired, because with liability
came presumptive authority. I say that I "won," but of course
I really didn't. The rabbis could not alter my plans, but they
had the right to stop construction, which they did. Then they
rigged up some plumbing and electricity, and moved without
permits into an unfinished building. No one was going to
stand in the way of ultra-Orthodox rabbis in the Old City,
not even Teddy Kollek. The yeshiva remains unfinished, yet
inhabited, to this day.

My early years in Jerusalem held one further disappointment: a failed plan for the Western Wall precinct. The demolition of the medieval Mughrabi Quarter, fronting the wall in an area known as the Valley of the Cheesemakers, had erased an entire neighborhood that now survived only in my childhood memories. I had known the area adjacent to the wall as an intimate space, a tiny clearing reached only after negotiating tight alleyways. The Herodian building stones in the wall were huge—six to ten feet wide or even larger. Some were weathered, some so smooth they might have been carved yesterday. Caper bushes grew from among the joints between the blocks. In the same small cracks, the faithful placed folded paper notes with handwritten prayers. The experience of standing alongside was at once amazing and mysterious. The wall seemed to rise into the heavens.

One of the first things that happened in 1967, right after the Six-Day War, had been the decision by Teddy Kollek and Moshe Dayan—the minister of defense—to physically clear the Mughrabi neighborhood in front of the wall. The people in the quarter were compensated, but they were forced to leave. The impetus for the decision was that, with the Old City now accessible again to Jews, and with the wall itself being a sacred religious and national site, the constricted approaches were inadequate. The razing of the neighborhood left a vast space, eventually paved, between the wall, at one end, and the Jewish Quarter, at the other—a space bigger than Manhattan's Times Square, but empty, undefined, and unattractive. The wall itself, once so imposing, now seemed diminished.

That was the situation when, in 1972, I was asked by a consortium that included the city of Jerusalem, the Department of Antiquities, and the Ministry of Religious Affairs—this last being the entity that had been given control over the area in front of the wall—to devise a plan for

My 1974 plan to expose the Western Wall to its Herodian base and create a terraced precinct sloping toward it.

the empty space. When I started thinking seriously about the possibilities, one idea that struck me forcefully was the need to re-create the massive presence of the wall—the high, looming reality I had known. The archaeologist in charge of the excavations in the vicinity, Meir Ben-Dov, with whom I worked closely, took me to see some of the shafts that Charles Warren, the British archaeologist, had excavated in 1867, indicating that the Herodian wall was in perfect condition for an additional thirty feet below the current surface, all the way down to the paving stones of the Herodian road that once ran alongside it.

What could be more logical, I thought, than to restore the Western Wall to its original height by digging down thirty feet and exposing the original base of the wall and the original street? The street would then become the surface where people would stand and pray—a street constructed of large, sturdy paving stones that had been part of the city during the Second Temple period. That raised the question of what to do with the plaza and how much of the adjacent expanse should be excavated as well. Digging the entire plaza to a depth of thirty feet would result in a cliff at the far end, where the Jewish Quarter begins, and might also run into a great many ancient structures, buried for eons. More relevantly, I suspected that the natural bedrock was rising toward the Jewish Quarter.

Explaining the Western Wall concept to Israeli president Ephraim Katzir, 1975. Teddy Kollek is on the right.

At some point, Ben-Dov asked me: What would you have done if you were Herod's architect? Working for Teddy Kollek, in a sense I was. I had been reading *The Jewish War*, by the Jewish Roman historian Flavius Josephus, and was steeped in his description of first-century Jerusalem. Eventually the solution hit me: just follow the bedrock. In other words, we should restore the area to something close to its likely configuration in Herod's time, with the city slowly rising in terraces, connected by stairways, from the wall and the original street upward to what is now the Jewish Quarter, on the escarpment. Designing the site this way turned it into a grand, amphitheater-like space. The Western Wall would regain its imposing scale because its full height would be exposed, re-creating a sense of intimacy for those standing next to it. The area closest to the wall would be reserved for prayer, while the public spaces on the rising terraces could accommodate as many as one hundred thousand people. The precinct could be a place for both religious festivals and national assembly.

That was the basic plan, and for a while there was enthusiasm from most quarters, including the religious establishment. I remember Louis Kahn voicing concerns when I presented the model to him and the rest of the Jerusalem Committee. He worried that the plan seemed too "exuberant." There were too many arches. He suggested "calmness." It may have seemed exuberant as one looked at the model from above, but I encouraged him to crouch down and see how very different the experience would feel at ground level. Yet his comment did give me second thoughts: perhaps the design *should* be calmed down.

The next few years proved arduous, and the plan for the wall was never realized. Design issues—exuberance versus calmness—were the least of the concerns. There was a long history of distrust between Israeli archaeologists and the religious establishment, particularly the rabbinate, which exercised power over the wall with growing fervor. The archaeologists sought to preserve significant finds as they were excavated, and the rabbis feared that the praying area would be compromised. Distrust deepened and proved unbridgeable. In retrospect, it was a mistake to have lodged so much authority at the outset in the hands of the Ministry of Religious Affairs, a mistake that can probably never be rectified, given the realities of Israel's coalition politics. As planning and discussions continued, the prime minister's office kept changing hands: from Golda Meir to Yitzhak Rabin to Menachem Begin—the lattermost in a government strongly dependent on the religious parties. This was not an atmosphere conducive to agreement on bold plans. Visit the Western Wall today, and you will see that the empty plaza looks the way it did five decades ago.

* * *

But the same cannot be said for the Jewish Quarter or for Mamilla, the area that lies outside the city walls beyond the Jaffa Gate—areas that were successfully transformed.

A structure in the Hosh District, in Jerusalem's Jewish Quarter, as it looked in 1967 (top) and after reconstruction, 1978.

My Jerusalem office was initially located in the Jewish Quarter, and from those premises we oversaw significant reconstruction and restoration work. I had opened the office largely with the encouragement of Yehuda Tamir, one of those influential figures who navigates skillfully between the public and private sectors. Tamir, a developer, headed the corporation set up by the government to rebuild the Jewish Quarter, and his insistence and ample persuasive powers drew me in.

Our responsibility extended to two specific areas: Block 38, located on the escarpment and overlooking the Western Wall, and the so-called Hosh District, an area abutting the Armenian Quarter. The work in the Jewish Quarter presented two different tasks. One was designing new infill buildings in cases where existing structures were a total ruin or beyond repair. The second was to save whatever could be rehabilitated, in the process often turning a single building into a structure with several individual apartments. These were mostly buildings from the Ottoman period—stone buildings with intricate vaults, domes, and arcades. Stripping and reconfiguring them was an extraordinary lesson, and I came away with a deep appreciation for the vernacular architecture of the city. At times I felt like a medical student with a cadaver, exploring the organs and sinews to understand the functional essence. With understanding came insight and inspiration. For instance, we reconceived the ancient domes for a modern era—not only making them transparent, so that they would glow gently from the inside at night, but also giving them a curved, retractable side that could open to the outdoors in good weather. Now, in the evening, looking down on the Old City from the Mount of Olives, you can see the glowing domes among the traditional ones.

As the buildings were completed, one by one, a lottery system was established to sell them to the public. There was a quota system for religious and secular Jews. The system also gave priority to those whose families had lived in the quarter before 1948. The objective was to achieve a balanced, integrated community.

At one point, I had a call from an entrepreneur named Yosef Golan. A number of larger structures in the quarter could not be divided into smaller apartments, and these had become available for individuals to restore. Joe Golan had bought a three-story courtyard house overlooking the wall

and wanted us to take charge of the restoration. As we did, I
became familiar with another ruin abutting Golan's house, a
structure in the same restore-it-yourself category. I bought
it and brought it back to life, and it has been my home in
Jerusalem ever since.

We started work on the Mamilla project in 1972, and it
was not completed until 2008. What had to be created in that
no-man's-land was a powerful link between the Old City and
the new city, between the ancient markets and the central
business district, between Arabs and
Jews, between religious and secular
neighborhoods. A mere park in that
area, as some had suggested, would
not suffice—a mere park would only
emphasize the separation. What was
needed was a mixed-use district whose
design was sympathetic to the terrain
and the surrounding architecture but
that provided places to live and work,
as well as hotels, shopping, and entertainment. The mission was to create a
district that could become a melting
pot and a place of meeting—an example of the public realm in the true and
full meaning of the term.

One lesson that a career in architecture teaches, or ought to teach, is
that designing a new building or a new
urban district is not a private discipline
—something that can come to fruition
in isolation, like a poem. The real world

Before 1967,
concrete
dividers west of
Jerusalem's Old
City separated
Israel and Jordan.

is all around you, with its many potential pitfalls, its evolving
social forces, its helpful and unhelpful personalities. Mamilla
offered a master class in all this.

Taking on a twenty-five-acre district in the heart of the
city, expropriating property, resettling residents and busi-
nesses, and then demolishing and rebuilding, would be a complex task under any circumstance. The "Jerusalem factor"—the
city's history and the associated emotional charge—made
this exponentially more complex. That said, the assignment
to create a master plan had a deeply personal meaning for
me. The area in the southern half of Mamilla, once known

An early sketch, in a 1972 letter to Michal, for transforming the former no-man's-land outside the Old City into a revitalized Mamilla district.

as the New Commercial Center, had been built in the 1920s by my great-uncle Eliyahu Shamah. He had created a district of industrial work spaces, offices, and shops facing the Old City walls in the hope of forming a vital business center that would attract both Jewish and Arab businesspeople. But in 1929, clashes between Muslims and Jews spread throughout the city. Hundreds were killed and the New Commercial Center was devastated. A year later, despondent, my great-uncle took his life.

The master plan we prepared for Mamilla, at the behest of the government and its development corporation, known as Karta, embodied the mixed-use vision. In essence, Mamilla is divided by the Valley of Hinnom. The valley bottom would be parkland. On the slopes of the valley, historic structures would be preserved, and new buildings on the slopes would be stepped, to accentuate and indeed exaggerate the natural topography. Throughout the Mamilla district there would be pedestrian thoroughfares and ample underground parking. Hotels and housing would bring residential and tourist life. Some vehicular streets would run belowground to get cars out of the way.

The parks and the shopping streets, together, would provide an active and diverse new center for the city. The strategic location would revive a tradition that had been characteristic of Jerusalem's walled city for hundreds of years: markets providing the place where all the city's communities could come together.

But there were difficulties from the start. For one thing, we had to cope with many changes of government, at both the city and the national level. In 1975, when the earliest plans were presented to the public, all hell broke loose. The Council for a Beautiful Israel, a public watchdog, opposed the plan, because it did not believe a contemporary development could be harmonious with the Old City. "Let it all be park," the society said. In this it was supported by the Society for the Protection of Nature, another public watchdog. Most seriously, an objection came from within city hall. The deputy mayor, Meron Benvenisti, an up-and-coming intellectual who had entered politics and become Teddy Kollek's key adviser, opposed the plan. He thought it was "too grand for Jerusalem" and that it would fail commercially and become a "white elephant." Benvenisti recruited David Kroyanker, from within the city planning department, and produced with him an alternative, minimalist plan for the district, with much of the area developed as park and pedestrian paths. The proposed development would not bridge the Old City and the new. It would send the wrong message—of a segregated Old City, a precious, museum-like space. In the end, Kollek reacted to the Benvenisti plan by issuing a statement reaffirming his support for ours. Benvenisti gave up the fight and resigned. I was sad to see him depart. I respected Meron and regretted that I had failed to convince him of the merits of what we had proposed.

Another interruption, much later, came in the form of the two intifadas—tragic outbreaks of sectarian violence that altered life in the entire region in countless ways. One consequence was that Mamilla's original developer, the British company Ladbroke Group, owners of Hilton International, withdrew from the project. Ladbroke was replaced by the Alrov Group, an Israeli real-estate company headed by Alfred Akirov, which led to a happy, decades-long collaboration that saw Mamilla through to completion. It was Uri Shetrit, my GSD student who returned to Israel and was now heading my office there, who suggested we approach Akirov. Today, Akirov regards Mamilla as his legacy, often remarking what a privilege had befallen him, to be able to help rebuild the heart of Jerusalem.

It sometimes seemed, though, that every time we sank a spade, we dug up trouble for Mamilla. Our plan called for the lowering of Jaffa Street, which fronts the Jaffa Gate, in order to allow pedestrian traffic at ground level to enter the Old City directly, without having to cross a busy thoroughfare. During the process of excavation, significant archaeological remnants were uncovered—a Herodian aqueduct, a Byzantine bathhouse, and much else. All of that had to be accommodated somehow. (We ended up building the submerged road on stilts, to make the archaeology accessible.) At another point, as bulldozers dug a parking garage, a cave was discovered containing more than five hundred skeletons. The remains were Christian—there was a Greek cross at the entrance to the cave—but, as always when human remains are found in Jerusalem, the Orthodox Jewish community descended on the site, arguing that some of the dead may have been Jewish, and brought construction to a halt. (The impasse was resolved only when the city government, in the dead of night, removed the remains and buried them in the Greek Orthodox cemetery.) The rabbinate also demanded that the parking garages be closed on Shabbat and that an entertainment center and cinemas be removed from the plan, on grounds of sacrilege. This dispute, which erupted just as Alrov took charge, wound through the courts for five years before being resolved in favor of the project; in the interim, prostitutes and drug dealers moved into half-finished buildings.

The story of many Mamilla buildings: numbered, dismantled, rebuilt.

The infrastructure needs of the new district—water, electricity, power, and parking—were extensive, and a number of historic buildings were in the way. After long negotiations with the Department of Antiquities, we were allowed to dismantle the buildings, construct the infrastructure, and then restore the buildings exactly as we had found them. Workers numbered each stone with black or red paint as they took the buildings apart. When the buildings were reassembled, you could often still see the numbers on the surfaces. On one of my inspection visits, I saw a worker scrubbing the stones to remove the paint. I insisted that the numbers

should stay, as an unexpected building ornament that tells
a story of preservation.

As Mamilla was nearing completion, the chief of police
of the Jerusalem District insisted that there must be security
checks at each of the entry points to the neighborhood.
Because there were fourteen such entry points—too much
to handle—the police proposed to fence off the entire district and leave only two entrances, where security forces
could keep watch. I knew that this would undermine the
goal of having a neighborhood open to all, as well as the
goal of making Mamilla an organic part of the city. In their
minds, the police seemed to be thinking of Mamilla as an
unusually large shopping mall; in truth, with its shops and
residences and streets, it was a classic urban neighborhood that demanded to be treated like one. It took some
doing, but the police chief agreed to abandon the unusual
security plans.

By 2008 the project was finally complete—and it turned
out to be worth the trouble. Mamilla was an immediate commercial and social success. "There is almost no chapter in
this project that is not controversial," observed a writer for

How the Mamilla
district looks now.

Haaretz in 2010, in an article that was itself not uncritical. "Yet the outcome is a public space that has transformed the city of Jerusalem. And among the regular visitors to Mamilla are Palestinians from East Jerusalem." Mamilla is a rare example of a planned public space that performs as anticipated, and it is among the few places in Jerusalem where Arabs and Jews enjoy the city together. For most residents of the city, imagining it today without Mamilla is impossible.

* * *

I have devoted a half century of my professional life to the city of Jerusalem. There was a moment after Teddy Kollek's departure when I entertained the idea of running for mayor of Jerusalem myself—something that would have been ill-advised. Running would have been an ordeal, and winning would have been worse. That I was even tempted is an indication of my frustration with misguided development, myopic and treacherous politics, and a lack of vision about how a truly united city might function. The failure of projects I had pursued left gaping holes, shortchanging the restoration and rejuvenation of the Old City and its perimeter. The prohibition against high-rise buildings in Jerusalem was broken under Ehud Olmert's mayorship, when the city planner, ironically, was my protégé Uri Shetrit. The breach began at some distance from the Old City, in western Jerusalem, with the spectacular, if ill-sited, bridge designed by Santiago Calatrava as the "entrance to the city" from the coastal highway. Not much of a span was actually needed there, but Calatrava created a great piece of sculpture; the 387-foot-tall pylon that supports the bridge is visible from miles away. Around this newly formed nucleus, with the Binyenei HaUma convention center and the central railway station, a new business center was conceived. Today, the Jerusalem master plan calls for a half dozen forty-story towers, and the plan is well on the way to completion.

Had I the power to issue decrees, I would stop all planned high-rise building in Jerusalem. I would also begin an emergency restoration process in the Old City, prioritizing the completion of its infrastructure, its subterranean wiring, its drainage. I would provide subsidized cable television and legislate away all the antennas, dishes, and other devices that mar the roofscape everywhere. I would standardize the design of a solar water heater, carefully shaped to blend in, and at the same time systematically eliminate the wide variety of unsightly water tanks, which are no longer needed. I would eliminate all the illegal construction, helter-skelter building projects of every kind, while guiding and, if necessary, subsidizing the replacement of what has already been built in cheap stucco or tin with carefully designed stone structures. I would complete the transportation, parking, and service infrastructure at the perimeter that is essential to saving and serving the Old City. In parallel, I would eliminate all the wild, illegal traffic and parking that now occurs within the city walls. The Old City has been a pedestrian precinct for all of its history. Parking can be built outside the walls, and I would provide for an efficient, modular, small-vehicle emergency and delivery service to substitute for what now is the infiltration of delivery vans into every corner of the Old City.

At various times I have proposed all of these things. I do not wish Jerusalem to become a precious museum city—like Venice or Florence—a static time capsule, captured in amber. And I am well aware that my dreams for the city are probably no match for the forces of the unfettered marketplace. Rather, what is likely to happen is that the current patterns will prevail. Even so, I am convinced that preserving the historic city—nourishing it and providing the infrastructure to thrive—is the only way to ensure that Jerusalem remains a vital economic component of the region. Taking this path would also show the way for other places that face the challenge of maintaining a precious heritage while remaining culturally and economically engaged with the modern world.

A photomontage
of the proposed
Columbus Center,
looking west
on Manhattan's
Fifty-Ninth Street,
1985.

CHAPTER FIVE

Private Jokes in Public Places

In 1972 and into 1973, I was commuting so often from Montreal to Israel that, in the days before frequent-flier benefits, Swiss Air would upgrade me to first class anyway. This travel routine did not help my marriage. Nina and I were drifting apart. Although Nina and the children would spend summers in Israel, during the school year we were often separated. On one of my trips to Israel, where I typically spent a week at a time, I met Michal Ronnen. She was the daughter of a friend, the Jerusalem artist Vera Ronnen. Michal was just coming off her military service, where she was a social-welfare officer stationed in the Sinai Peninsula. I was in my early thirties. She was twenty. It still amazes me what great responsibilities young Israelis are given—and undertake—in the Israel Defense Forces. Michal was mature, warm, idealistic, and compassionate. She, too, had been on the cover of *Newsweek*—selected to represent Israel for an issue celebrating the nation's twenty-fifth anniversary—so she and I both shared that coincidental distinction. Our friendship became intense and intimate.

This is not the place to get into why and how my relationship with Nina fell apart. She did become aware of what was happening, and we lived with this awareness for several years. I felt deeply torn and guilty over what seemed an inevitability—that I would separate from Nina and embark on a lifelong relationship with Michal, as indeed I have. The trauma of Nina's past, notably her horrific experiences during the war, deepened the pain. But most of all, my concern was for our two children—Taal, who was then twelve, and Oren, who was eight.

* * *

In October 1973—a year after Richard Nixon's historic trip to Beijing—Canadian prime minister Pierre Trudeau announced that he would be visiting China to mark the opening of diplomatic relations between Canada and the People's Republic. Arthur Erickson and his partner, Francisco Kripacz, were invited by Trudeau to join the official delegation. In turn, Arthur and Francisco suggested that Nina and I join the trip as well. The four of us decided to set out ahead of the official delegation, traveling first to the Soviet Union, to visit Moscow and what was then called Leningrad, and then venturing on to Armenia. From there we flew to Siberia and rode the Trans-Siberian Express from Irkutsk through Mongolia to Beijing. We were on the train for three days, basically incommunicado. Disembarking in Beijing, we were greeted by a Canadian embassy official. When we asked, just to make conversation, if there'd been any news, he replied, "Nothing new today. Neither the Egyptians nor Israelis changed position." That is how I learned about the Yom Kippur War, which had broken out while we were crossing Mongolia.

Prime Minister Pierre Trudeau and his wife, Margaret, arrive in Beijing, October 1973.

That afternoon I managed to borrow a shortwave radio from a *Wall Street Journal* reporter and was able to listen to the BBC, which became my prime source of information for the next few weeks. Michal, who had been visiting friends in Copenhagen, had returned to Israel immediately and was drafted into the reserves and shipped to the Sinai, where she served on the front lines with a tank brigade. Her job was to document the relevant details concerning the dead and wounded and to communicate with the soldiers' families— a profoundly wrenching task. Her messages, sent via the Canadian embassy, kept me intermittently informed. As an Israeli in Beijing at a time when communications were not what they are now, I often felt like I was on another planet, disconnected from the world. One of Michal's missives in particular was memorable:

> *I sat a long hour with the deputy commander of our unit that was hit on that terrible day—the 24th, when we entered the Suez Canal. A bullet penetrated the left side of his brain. He is totally paralyzed on his right side, and his memory and speech are just beginning to return now. He is missing words, and I had to hold back my tears when for many long moments he tried to tell me something and was not able to remember the word or the subject, and he kept saying "Noo Noo Noo." He is 47 years old, handsome, and very sympathetic. I was shocked when I saw a blue number tattooed on his arm from the camps.*

Little moments underscored the reality. As we arrived one evening at the Great Hall of the People for a state dinner, we could see Arab delegates embracing and congratulating each other on Israel's downfall—this was day three of the war, when the situation looked very bleak for Israel. My knowledge of Arabic, familiar since childhood, made it painful to hear these victory celebrations, which turned out to be premature. Traveling on official business with Arthur, Francisco, and Nina, as well as the prime minister; meeting with the Chinese leadership, including Zhou Enlai; complying with a packed schedule; worrying about the war; worrying about Michal; wishing I could be serving in Israel—it was an intense period. The experience of being on the ground in China in such a highly choreographed fashion only heightened the sense of confusion; the lives of the Chinese people had clearly improved, but as I would write to Michal, the regime was "manipulating, controlling, causing suffering . . . as the children sing their happy songs." Visiting China in 1973 was among the most fascinating and informative trips I've ever experienced; at the same time, I was in constant turmoil.

We traveled through China for four weeks, ending up finally in Hong Kong. Nina flew east to rejoin our children in Montreal. Impelled by memories of my absence in 1967, I flew west to Israel. In some ways, I would say, at that point, with respect to my marriage, the die was cast. In Israel I was immediately enlisted into the reserves, becoming part of the education corps—created to bring cultural and other fare to the troops. I was told that I would be flying to Egypt, across the Suez Canal, to an Israeli redoubt near Ismailia.

Members of the Architecture Society of China, along with Nina, Arthur Erickson, Francisco Kripacz, and myself, Beijing, 1973.

The location of a major battle, the site had been given the name "the Chinese Farm." (The area had held an experimental agricultural facility and had used equipment imported from Japan; Israeli mapmakers saw the Japanese characters, thought they were Chinese, and gave the location a new name.) A cease-fire was by now in place. I brought with me a carousel of slides about the work we had done in planning for the future of Jerusalem, but I had also developed photographs from China and edited them into a slideshow. It opened with photographs of Zhou Enlai welcoming the Trudeaus at the airport in Beijing.

Thinking back, 1973 was a time when few in the West had any knowledge or experience of China. I had a projector that could be plugged into the battery of a tank, and as I moved from one unit to another, I offered the soldiers a choice—Jerusalem or China. China won in most cases. At some point, Michal was able to visit. I remember a soldier coming to tell me that an officer was asking for me. Private Moshe then met Second Lieutenant Michal at the camp gate.

* * *

During the 1970s, more than a third of my time was spent in the air or at work in far-flung destinations. I often took either Taal or Oren, then in their teens, with me on business trips. On location, I would sometimes bring them to meetings, sometimes to project sites. These were special moments, the circumstances often fascinating. It was hardly a substitute for full-time fathering, but it helped. The trips are etched in their memories to this day.

The firm's principal office, then still in Montreal, was getting work overseas but virtually nothing in Canada, and years would pass before it did—not until after I had relocated it to the Boston area. I can only speculate about the reasons for this dearth. Que-

As a private during basic training for the Israel Defense Forces, 1973.

bec was in the midst of its separatist project, and local work favored French-Canadian firms. Private developers, both in Montreal and Toronto, seemed uninterested in working with an architect who might, given the example of Habitat '67, prove too experimental and risky to take on. Whatever the reasons, the firm in these early days always seemed to be on the edge financially. Fees from the ample work in Israel did not arrive speedily; the country has what some have called a "poor payment culture." We often borrowed from banks to tide us over.

My recollections of these days recently became more vivid when Michal came across boxes of my letters to her in the basement of our home. She had been living in Israel at the time. In March of 1973, I described some of my professional unease as I took on a particular commercial project back home:

> *What really troubles me, my Michal, is that I feel that I am being swept by a current; that in my desire and frustration to get something built, I'm taking on work that is problematic, doing my best to make it acceptable, digestible. But the reality is that this project has a regular commercial framework, and it would be almost impossible to uplift it above its inherent constraints or limitations. In other words, am I doing all of this because I'm desperate to build and I need work to sustain my office, or because what I'm doing is going to be good? It's not an easy question and I have been quite tormented by it.*

A year later, I wrote to her from Montreal about the sudden death of Louis Kahn:

> *I called Anne Tyng in Philadelphia and learned the sad details. On Sunday, Kahn left India for Philadelphia. He missed his connecting flight in London and arrived late in New York. Because of the delay, he missed the connection to Philadelphia. Apparently, he went to Pennsylvania Station, where he went to the men's washroom, and there suffered a fatal heart attack. He was found dead at 8:45 in the evening by the police. They did not identify who he was. He was taken to the morgue, and then a routine message was sent to the Philadelphia police that "One Louis Kahn was found dead in the station." On Monday morning, as Lou did not arrive at work, they began searching for him around the world. Eventually, the Philadelphia police figured it out and informed his wife and the office of his death, all of which did not happen until two days after he was discovered. I feel very*

sad. I received so much from Kahn. My work in his office was a very important time in my life. I learned a great deal from him, even though I was a rebellious student.

Louis Kahn and his family—or, more accurately, his families—continued to have a significant presence in my life. Besides Anne and her daughter, Alex, there was Nathaniel, born when I was working in Kahn's office in 1962, the son of Kahn and the landscape architect Harriet Pattison. In January 2000, I received a call from Nathaniel inviting me to join him in a filming session for his frank and discerning documentary about his father, *My Architect*. He suggested that we meet in Israel and retrace Kahn's steps on a trip he had once taken from Jerusalem to the Judaean Desert and the fifth-century Mar Saba monastery. And we did.

A settlement of Inuit "Matchbox" dwellings, 1974.

A week after my letter to Michal about Kahn's death, I was heading north—I had received a commission to design housing for the Inuit community in Canada's Arctic. Back then, the entire Canadian north was under the jurisdiction of the government of the Northwest Territories. It oversaw, colonial-style, the welfare of the indigenous First Peoples: Inuit in the northeast and other nations in the west. The Inuit had slowly evolved from a fully nomadic existence to life in settlements. "Matchbox home" was the term used to describe the tiny, foundationless wood residences that dotted the tundra, well above the tree line.

Better, more suitable houses in planned villages were needed. As all building materials had to be imported from the south, and the northern seas were navigable only in the summer, here was a perfect case for prefabrication. In preparation, I toured several Inuit communities, flying from improvised runway to improvised runway in small, short-takeoff Twin Otter aircraft. Daylight was perpetual in summertime when we began the project. On the ground, whiteouts would sometimes envelop us, the air condensing so thickly that you could not see your hand in front of your face. Later, in winter, I would experience the oppressive gloom of perpetual darkness. My hope was to be able to make the houses glow, like

igloos. In another of those letters to Michal, this one date-
lined "Edmonton to Yellowknife," I drew sketches of the
clothing I saw and of ideas for Arctic housing as I pursued
the Inuit project:

Sketch of Inuit
parkas from a
letter to Michal,
1974.

> *Everything, everywhere is white. I started my journey to*
> *the North. In two hours (1500km) I will be in Yellowknife.*
> *Already at Edmonton's small airport, I felt I am entering a*
> *different culture. Everyone was wearing furry parkas and deep*
> *hoods; mostly they were decoratively embroidered. Even my*
> *own parka, which I purchased before departure in Montreal,*
> *has embroidery on the sleeves and trim. . . . As it always is in*
> *a new place, I feel that I am your eyes, [and that] I see also for*
> *you, absorbing, understanding, and remembering so that I can*
> *describe and share with you everything.*

Initially I had assumed, given the cold, that the houses
might be linked to each other, as well as to the school and
other public buildings, but meetings with the Inuit community made it clear that, no matter how cold, getting into the
outside air was a priority for them. We designed the houses as
octagonal spaces, with multiple sleeping rooms surrounding
a central domed living space. The walls were prefabricated,
stressed-skin plywood panels insulated with white fiberglass,
and the dome was also of fiberglass, and translucent. Concrete
piles driven into the permafrost would support the homes
off the ground.

As I got to know the Inuit community, I became its advocate. But the community was not our client. The bureaucrats
of the Northwest Territories were the
clients, and they believed we were raising the bar too high, creating expecta-
tions that could not be sustained for
indigenous communities throughout
the north. In the end, the government
decided not to proceed with our plan.
The Northwest Territories built a few
more schools and community buildings, but the Matchbox houses prevailed for another generation.

Design for a
prefabricated Inuit
home, 1976

In 1977, as our Arctic efforts wound down, a new client
flew into Montreal from Singapore on his private jet. Robin

Loh was an Indonesian Chinese shipbuilder (and former taxi driver), who had settled in Singapore and was now involved in real estate. Loh knew about Habitat and was particularly interested in the idea of modular prefabrication. There had been a slump in the shipping business, and he wanted to convert part of a shipping yard into a factory that made housing modules that he would ship to the Persian Gulf states and elsewhere. He invited me out to Singapore. I could not have anticipated that the island city-state would become a major center for my work for decades to come.

Loh was my first introduction to a self-made Asian tycoon. In Singapore, I met his vast extended family, all somehow incorporated into the business. I also came to realize how extensive the business was—shipping, real estate, palm oil, and other ventures. Once, at his office in the Robina Tower—named for himself—an early high-rise in downtown Singapore, Loh opened a concealed walk-in safe and started bringing items out to show. A solid gold dinner service once owned by King Farouk. A Stradivarius violin.

Preliminary sketch for Robin Loh's "integrated resort," to be built in Gold Coast, Queensland, Australia, 1981.

Loh had secured two prime sites for development. The first was in the heart of downtown, at the top of Orchard Road, which would obviously provide luxury housing. The second was in the Tampines area, near Changi Airport, where Loh hoped to build affordable housing. With Loh, I also traveled to Australia, where he had secured thousands of acres in Queensland to create a new urban sector—to be called Robina, again—in the city of Gold Coast. It would feature housing and a commercial district but also something he called an "integrated resort": a combination casino, convention center, and shopping and entertainment complex. This was my first encounter with a term and concept that would define a significant part of my work later on, though not in collaboration with Loh. Only one project with Loh actually came to fruition—the Orchard Road development, where we built a pair of thirty-story luxury towers. The resort in Australia, where I had envisaged a complex rising out of an artificial lake, ran into opposition; in the late 1970s, the prospect of an Indonesian Chinese Singaporean outsider getting a license to operate an integrated resort was too much for Australian authorities to contemplate. But the association with Loh established our firm in Singapore and taught me about a place and a culture in which I would one day be spending much of my time.

Finished model of
the Queensland
resort complex,
1981.

* * *

I t is hard to predict where projects will come from. For
instance, my relationships in Israel led me, counterin-
tuitively, to Iran and Senegal. Joe Golan, whose house abutting mine in Jerusalem I restored in the early 1970s, was like
a character from a John le Carré novel, one of those people
who roamed the world and seemed to know major politicians
and business operatives everywhere. His knowledge of Arab
countries was immense, and behind the scenes he facilitated
the emigration of Jews to Israel from several of those countries.
Golan had a big network in Africa and also in Iran, prior to
the 1979 Iranian revolution. Shah Reza Pahlavi was in power
then, and his wife, the Shahbanu, Farah Pahlavi, was a glamorous and influential global figure. I myself was starting to get
connected in Iran. I had attended the Persepolis International
Architecture Conference that the Shahbanu had convened,
in 1974, where major figures in architecture had gathered—I.
M. Pei, Paolo Soleri, Bruno Zevi, James Stirling, Jørn Utzon,
Arthur Erickson. I met the Shahbanu and her architectural
adviser, Manouchehr Iranpour, which led to discussions about
creating a Habitat in Tehran, among other projects. In 1976,
I wrote to Michal about some of my impressions. Bahadori
was Karim Pasha Bahadori, a close adviser to the Shahbanu:

> I had dinner with Manouchehr Iranpour who is the court
> architect of the empress. He is a Zoroastrian, strange character;
> giving me presents, taking me out for elaborate dinners, insisting
> that I have more caviar. If I don't consume at least a half a bottle

of vodka his feelings are hurt. The following day we started serious work. I presented the Habitat Tehran drawings to Bahadori, who was truly excited. I asked him if we have an audience to present the design to Her Majesty. He said that this was very difficult as they were leaving on Friday, but he will try. Towards evening there was a message in the hotel. The Empress will see me at the palace at 2:15 in the afternoon the following day.

At 1 pm they fetched me to the palace (blue suit, turquoise tie). We entered through the palace gate. A gentleman covered with medals came forward to receive us, and from there, by foot, we walked along one of the most beautiful gardens I've ever seen, an alley of trees, beautiful lawns, cascading water ponds, and the song of birds filling the air. We walked through the garden for 200 meters and arrived at the white palace. From there we entered the ground floor, decorated in red and velvet, and after a ten-minute pause, I was invited upstairs to the piano nobile. As we entered, we saw a dozen Louis Quatorze chairs arranged in a circle, our renderings and plans set on them. The model was displayed in the center, and my books on a separate table. The empress entered. She was tanned, wearing a short silk skirt with a green blouse.

I presented the plans. We spoke for about an hour. She was enthusiastic about the design for Habitat Tehran. At some point she said, "So what is next?" Bahadori responded, "We have been waiting to hear whether the project is approved by Your Majesty, and now we will proceed to implementation." Her Majesty responded, "Approval is hereby granted."

In these letters I also conveyed to Michal my ambivalence about commissions offered by autocratic governments. At this very same moment, I was discussing the possibility of redeveloping an area of Manila to serve its poor residents, and had been invited to visit the Philippines by Imelda Marcos. Her exhibitionistic consumption—she famously owned three thousand pairs of shoes—along with that of her friends, was repellent. "What should my position be?" I wrote. "Not to work for any regime with whose governance I have a disagreement? Where is the separating line? Italy is good, the Philippines not? Is my contribution as an architect to the regime or the population?" That last question is the key one. Who benefits? Ordinary people, or just some pampered elite?

* * *

J oe Golan operated at a very high level, pulling opportunities out of thin air, and the idea he had in mind would
improve the lives of many. He envisioned a grand international
scheme involving Iran and Senegal—a country to which he
actually held a passport. Senegal's president, Léopold Senghor,
a close friend of Golan's, had given it to him after Golda
Meir momentarily revoked Golan's Israeli passport on the
grounds that his wide network in the Middle East marked
him as a man who couldn't be trusted. I remember Golda
as tough and resolute, for all the softer depictions of her in
the American press. When I presented her with my plans
for the Western Wall, there was a sternness in her manner
and even her stance. Her legs were planted to the floor like
Doric columns.

Golan's scheme was essentially a barter deal: the Iranians
would supply oil to Senegal, and the tankers that brought
the oil would return to Iran with phosphates for fertilizer.
But there was more: the Iranians would also finance the
building of a new city in Senegal, which would be called
Keur Farah Pahlavi, after the empress. It would include
a deepwater port and a refinery, and would be connected
by rail to the phosphate mines. It would be built along the
Atlantic coast, a hundred miles north of Dakar, Senegal's
capital. The new city was a joint Iranian-Senegalese project
in which the Shahbanu was personally invested. So were
the leaders of Senegal—President Léopold Senghor and his
prime minister, Abdou Diouf. I was given the commission
to design the new city.

A special committee was formed that included the prime
minister of Iran, Amir-Abbas Hoveyda; the oil minister, Reza
Fallah; and the Iranian sculptor Parviz Tanavoli. I started
shuttling between Tehran, Senegal, and Montreal. Michal
was often with me.

In terms of urban planning, this was an astonishing
opportunity, though the issues in Senegal were far removed
from those of the dense Western cities I was familiar with.
The new city, which was planned for one hundred thousand
people, would be spread out. There would be no high-rise
buildings. Most Senegalese live in compounds with extended
family—men have multiple wives, and there might be three or

four structures within a walled compound. There is almost a village-like scale to urban arrangements in Senegal. I found dealing with this new world to be very exciting.

In general terms, the site had already been selected, but the question to be answered was exactly where, on the vast site, the city would go. The topography was varied. To the west was the Atlantic coastline, long and uninterrupted, backed by rising sand dunes a half mile deep. The land then dropped to a lower level, where the dunes blocked the drainage to the sea, creating a series of natural ponds and lakes within a ribbon of fertile land. The terrain then gently rose toward a series of low hills. The port, of course, had to be on the coast. I decided to place the city behind the dunes, away from the coast, with access roads crossing the dunes to the sea in two places, one for the port and one for recreation and access to beaches. As someone who had grown up in Haifa and enjoyed the sea, my initial intuition had been to build the city along the beaches on the coastline—Haifa, Tel Aviv, and Casablanca share this character. But the entire coast was largely free of human settlement; even the fishermen would build only temporary lodging on the coast, retreating to permanent villages behind the dunes. Clearly the coast was inhospitable.

In my plan, the linear series of lakes and ponds behind the dunes, running parallel to the coast, would be incorporated into the spine of the city—a mixed-use zone of shops, offices, and public buildings. Residential neighborhoods would straddle the spine on both sides, extending toward the dunes in one direction, to the foothills in the other.

Along with the overall site plan came the designs for a variety of housing types, schools, and public buildings. They would be made using local materials and local methods, reflecting the local and particular lifestyle. This exercise brought home the extent to which architecture, particularly housing and urbanism, is not exportable (or importable). One has to begin by understanding how people in a certain place actually live—their traditions, their economy, their technology—which may be very different from how people live five thousand or even five hundred miles away.

It was never lost on me that we were in a Muslim country, and I was an Israeli playing a pivotal role—but this never became an issue. I developed a friendship with President

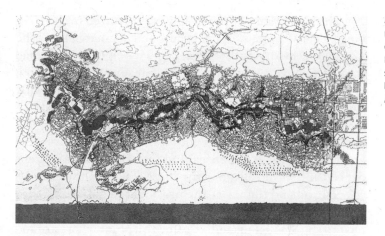

Aerial plan of the new Senegalese city Keur Fahrah Pahlavi. The dark strip along the bottom is the Atlantic Ocean.

Senghor, a renowned poet. He was exuberant, cheerful, enthusiastic. Physically he was not tall, but he had a very solid frame. French was the language of his poetry, and he would become the first African man elected to the Académie Française. The prime minister, Abdou Diouf, was a figure out of a movie: tall, slender, proud—one could imagine his willowy shadow moving serenely across the desert. The two, Senghor and Diouf, made quite a pair—contrasting and yet complementary. Senghor made me promise to learn French as a condition of taking the job, a promise I never quite managed to honor. But the conversations and correspondence I had with him were very much out of the ordinary. At one point we discussed the way in which building forms might mimic the laws of nature, and I found myself writing a philosophical letter—not about the project itself but about architecture, nature, and design—in which I made some specific points about nature that had caught my attention:

**The nautilus shell is a logarithmic spiral which is the result of the animal's need to have its house and body grow continuously and proportionately—the only mathematical possibility of such growth.

**The bone structure of the wing of a vulture is the most magnificent network of three-dimensional lattice giving the maximum strength for minimum weight.

**The geometric network of the bee's hive is the means of storing the most honey with the least wax.

The Shahbanu was personally very involved in the planning. She was famous for her beauty and her manner, but she was also accessible and a listener, and it helped that she had studied architecture. In 1978, we were ready to break ground. A lavish ceremonial plan called for festivities in Dakar that would then roll by motorcade to the site of the planned city. A sculpture by Tanavoli would mark the cornerstone. The Shahbanu arrived with her entourage. The highways were lined with the flags of Senegal and Iran. Senegalese citizens danced and waved all along the route. That night there was a reception and dinner at the presidential palace. Toward the end of the evening, as guests were leaving, I was walking next to the Shahbanu, and we came to a point where a grand stairway descended two or three stories. As we started down, the heel of her high-heeled shoe snapped. She grabbed my arm and, without saying a word, continued imperceptibly, with one foot in heels and the other on tiptoe.

She would need such poise in the months ahead. A few weeks after the ceremony, Iran was engulfed in its Islamic revolution. Ayatollah Khomeini returned from exile. The Shah and the Shahbanu fled. Iran was convulsed for years. In the eyes of history, it is only a footnote, but one of the first things Khomeini did was cancel the Senegal project, even though the idea that had given it life—oil for phosphates—still made sense for both countries. It was a relic of a past regime. There was nothing to be done. Needless to say, Habitat Tehran was also not to be.

* * *

People often ask: Given the many unbuilt projects in an architect's life, what happens to all the designs? A considerable body of unbuilt work is inevitable for any architect who seeks to operate in the forefront of the design world. About half the designs we have developed remain unrealized. The stages these designs have reached vary widely. It is most painful when these unbuilt projects have gone beyond the concept phase, and we have spent one, two, or three years developing the design in great detail, and then face the disappointment of cancellation for one reason or another. The financial and emotional consequences can be severe. There are sometimes—rarely—competitions we enter together with developers, where we are fully compensated for all expenses, regardless of outcome. Jewel Changi would be an example, although the issue was moot: we won the competition. But most competitions offer only symbolic compensation.

The economics of an architecture office such as ours are unforgiving. We're not a corporation with an organized marketing department, nor could we ever be. For us, "marketing" consists of submitting proposals either for competitions or upon invitation from a particular client. In other words, marketing amounts to doing the work itself. We know from long experience that the cost of entering a competition will typically run to double or triple any compensation that may be offered. For a major project, the competition stipend might be $250,000; we will easily spend $750,000 or more. Some firms are able to cut corners and work quickly—producing renderings without going through the whole process of site analysis and program analysis and understanding the totality of what is involved. I don't know how to do that. When we present something as an idea, it's with an intention to build; we have done enough work to know that the design will stand up, that it's structurally sound, that it's buildable, that the concept will hold water when we come to develop it. That takes time, and it takes thought. It is a condensation of the normal design process we would pursue for any client. Compress it as much as we can, and it will still take seven or eight people two or three months.

The bigger dilemma—for the profession as a whole, not just for our office—is the unending tension between fixed numbers and variable conditions, and, additionally, between

At the groundbreaking of the planned city with President Léopold Senghor and Empress Farah Pahlavi, 1979. The prime minister of Iran; Amir-Abbas Hoveyda, is to Senghor's left,

competitive fees and a maximum quality of service. Fixing the fee means fixing the fee for a project that will extend out over four or five years, from design work in the office to construction on the site. One has to project through all the different phases and anticipate all the different conditions. The variables are endless: unexpected variations in materials and methods; schedules that get stretched; clients who change their minds about this and that—something they're supposed to pay for, though they always resist. This problem of risk is inherent in the profession. If one tries to deal with it by providing a bigger cushion in the fee, then one becomes less competitive relative to the fees proposed by other firms. Clients may say sincerely that they simply want the best architect, but when fees vary significantly, the variation makes a difference.

If everyone we were competing against offered the same quality of service we do, the decisions made by clients would at least involve a comparison of apples to apples. But often we're competing with more commercial firms whose level of service is fundamentally different from ours; fewer person-hours spent on design, fewer person-hours spent on supervision at the site. One can try to explain all this, and one does, but it takes time, and clients may not always appreciate what they will be missing.

Even when we get the commission, the fixed fee is always in tension with the desire to do more. We are ambitious about the end product. We want perfection, which cannot always be reconciled with prudent business practices. Let's say we've worked on a design for two months, and suddenly there's a eureka moment: a new concept, something that comes from the team, not from the client, who happens to be satisfied with the design as it is. Exploring the new concept, which may take weeks, is a cost we have to bear ourselves. Meanwhile, the clock is ticking. The finance people in the office are screaming.

In the end, some projects never get off the ground. What unbuilt projects can do is help one generate an inventory of ideas. Many years ago, we designed a complex—a kind of getaway conference center—for Clal, a Jewish educational institution, on a rural site in New York State. The site contained a stream running through it, and we proposed to dam the stream, creating a pond, with the building constructed over the dam and forming a causeway from one bank to the other. The project never came to fruition, but the central idea was revived and transformed when we were commissioned to create a campus for the Superconducting Super Collider in Waxahachie, Texas. That project foundered when Congress halted funding for the supercollider effort altogether. But the fundamental design concept remained a promising one—and when I was asked to conceive the Crystal Bridges Museum, in Arkansas, it provided inspiration for a project that did indeed come spectacularly to life. Our project in Senegal may have come to naught, and yet it generated ideas that took root in the even larger project for the city of Modi'in, in Israel. I can think of many more examples from my own work. Other architects have examples from theirs. Good ideas are rare. No one wants to leave them moldering.

There is much about the architect-selection process today that is unpredictable. Some projects are so ambitious and enticing, and yet reimbursement for participation in the process is so negligible, that competition can prove both tempting and costly. Selection committees can sometimes amount to a visionless assembly of contradictory interests. All this is true and often leads to a dead end. But the ideas and insights from some "dead ends" can be incubated until the time is right.

It is also true that one thing leads to another. The involvement with Iran may have been unexpectedly curtailed, but I made many friends in the process. Among them were the German artist Karl Schlamminger and his wife, Nasrin, an Iranian whom he had met at art school in Istanbul. Karl had converted to Islam and moved to Iran. I will never forget being taken by Karl and Nasrin to the music academy in Tehran, with Michal, to hear a master of the *ney*—the Iranian wood flute. I can still hear the melody. Following the revolution, we collaborated with Karl on various projects. He was without parallel when it came to incorporating Arabic calligraphy into modern design. His clock tower marks the hours at the Harvard Business School, outside the chapel I would one day design.

Another friendship formed at this time was with one of Iran's prominent architects, Nader Ardalan. An Iranian of Kurdish heritage, from a distinguished family, Nader was then married to Laleh Bakhtiar, a scholar of Sufism. His mustache and skin tone often led people to think we were brothers. Nader had studied at Harvard and worked in the United States for many years before returning to Tehran, where we met, and establishing his own thriving practice. In 1977, he decided to open an office in Cambridge, Massachusetts, an American practice in parallel with the main office in Tehran. Little did he know that, as the Iranian revolution took hold, his Cambridge office would become his only office.

Nader moved to Boston and bought a house in Brookline. He also became involved in the Graduate School of Design at Harvard. One day, in 1978, the dean of the school, Jerry McCue, mentioned to him that Willo von Moltke, who headed the urban design program, was retiring, and the program would be looking for a new director. My name had come up, but the dean had dismissed the idea, thinking I was too busy. Nader was aware that I was getting anxious in Montreal, that for several years I hadn't gotten any local commissions, and that my personal circumstances were in flux—Michal was part of my life, and I had separated from Nina. He told the dean not to be so sure.

Nader broached the idea with me, and it found fertile soil. I said I'd have to move my office, as I couldn't continue to run my practice and commute at the same time from Montreal to Cambridge. He encouraged me to make the move. My interview at the Graduate School of Design, in December, coincided with the famous blizzard of 1978— quite an introduction to Boston, which remained immobilized for a full week. I wrote candidly to Michal when the visit had ended.

> *By a miracle I made the last flight this morning before they closed down the airport because of the weather. It's been a full day, endless meetings at Harvard with the dean, with the chairman of the department, with the senior professors and the young ones as well. They're checking me out and I'm checking them. Jerzy Sołtan, the old veteran Polish professor from the days of Le Corbusier, told me that I must absolutely come, that the profession is in danger of those who treat architecture as purely visual, as a painting. The academic atmosphere, however, weighs on me. I feel the bureaucracy and tensions between the faculty. I feel how all this could suck the energy*

out of your marrow. I also met with the urban design students.
I gave them thirty minutes of the essence of my thoughts, and
heard them out; it was productive.

In the end, I decided that Boston could prove to be a
better base for my work. It was not an easy decision. Montreal had been home since our family left Israel a quarter of a
century before. My daughter Taal was by then starting college,
but Oren was still in his teens; making visits and taking trips
together cannot compensate for absence.

Michal came to Boston. For the first time, we would live
permanently in the same place. Michal's photographic work
has long been bound up intimately with my architectural work.
From a single vantage point in our home in Jerusalem, overlooking both a narrow alley and a wide view of the Western
Wall, the mosques atop the Temple Mount, and the Church
of the Holy Sepulchre, Michal would eventually produce a
photographic record that has been published in folio form
with the title *Under My Window*. In Israel, Michal had studied
social work at the Hebrew University. When we arrived in
Boston, she did graduate study at Brandeis University and
earned a master's degree in sociology. As we traveled together,
she often photographed our buildings, and discovering that
she was good at it, she became the photographer of the architectural models in the office. Then she embarked on her own
journey in photography—not only the Western Wall but also
anthropomorphic trees, vapor trails, the genocide trials in
Rwanda, and more. Her keen photographer's eye has made
her a valuable critic. Often, when I am traveling, and a presentation is being prepared at the office—photos, renderings,
models—I ask her to drop in and check things out. Of course,
her involvement goes well beyond that; it must not be easy
to be an architect's spouse. I come home every evening and
unload the many issues that have bedeviled the day. Michal
is the trusted counselor.

Relocating the firm to Boston was not an easy process. I
could not bring everyone, and not everyone wanted to come.
There were bureaucratic hurdles—even back then, acquiring
green cards took work. In the end, several senior members
of the team—Bill Gillit, Isaac Franco, and Philip Matthews—
made the move south, to the somewhat balmier precincts of
Boston. We found space in Faneuil Hall Marketplace, which

had just opened, right next to Ardalan Nader's own office. I started teaching, running the department, and building up a practice in Boston.

At Harvard, I began with a top-to-bottom review of the entire curriculum. I invited the faculty to present new ideas and to be critical of the way we did things. We introduced something called the development studio, where architects and landscape architects who were pursuing an urban-design degree would do a studio designing an actual urban sector, and where the faculty would consist not just of architects but of developers. The studio would also have an economics and business component. The idea was to immerse students in the real-world issues they would face when taking on any large-scale development, even the transformation of whole neighborhoods: the financial constraints, the building codes, the public presentations, the sometimes vicious (even if invisible) city politics. This approach was revolutionary, because the teaching of architecture in academia tended to focus on design and to avoid the financial and development issues of the actual world—which is, of course, the world that every architect and urban designer has to face as soon as he or she graduates.

An image from Michal's *Under My Window* (2018), a book of photographs of the Old City of Jerusalem from a single vantage point in our home.

The task of designing a project and seeing it through to the end is shot through with financial implications. The more artisanal the firm, the more any individual architect needs to know. A megafirm may have a finance department the size of my entire staff. At some such places, the finance department seems to roll forward on its own, noting every "change order" the way a hospital charges for every aspirin tablet. But at a firm such as our own, each project manager needs a balance of skills. She or he takes responsibility and must continually ensure that excellence and profitability are compatible.

I also realized that, if we wanted to do a series of exciting studios in a place where context was central, we should think about doing them someplace other than Boston, which had become the context more or less out of habit. We had students from every part of the world. My first class had participants from Venezuela, Spain, Japan, Israel, Canada, and Singapore, in addition to those from the United States. Because I was deeply involved with Jerusalem, I thought it would be exciting to do a series of urban-design studios there and set them up in such a way that the local communities and city authorities could have input.

With Teddy Kollek's help, I found a donor, Irving Schneider, a New York developer and philanthropist whose daughter Lynn was studying architecture at Harvard. Schneider was committed to Jerusalem. Over the course of several years, he gave the Graduate School of Design well over $1 million. Some 150 students would participate in the program. The Harvard-Jerusalem venture became the first of the so-called sponsored studios at the GSD, a format that is now commonplace.

We traveled to Jerusalem together—a mix of architecture, urban-design, and landscape students, plus members of the faculty. The students met representatives of the Palestinian community as well as Israelis. They met with scholars from many disciplines and with people in the municipal and central governments, and out of these discussions came stimulating ideas. The students were tasked by the faculty with developing plans for almost every sensitive part of Jerusalem: the Damascus Gate area, the Mamilla area, the various quarters of the Old City, the landscape corridors that lead to Bethlehem and elsewhere. To illustrate the complexities involved, one element of the Damascus Gate plan called for a new bus terminal that would serve both the Israeli bus company, Egged, and a dozen or so existing Arab bus companies. The political challenges of building such a terminal were as significant as the design challenges. Consider a simple question: Would Israeli and Arab companies share all facilities in common? (Not a simple question, it turned out.) After four years of these exercises, two of my students—Uri Shetrit and Rudy Barton—and I collected the group's ideas into a

With our friend Nader Ardalan in Isfahan, Iran, 1978; he helped bring me to Boston.

book called *The Harvard Jerusalem Studio*. In one place we noted pointedly:

> *We felt that Jerusalem studies should not avoid political questions; they are an integral part of the planning process. No planning decision can be made in Jerusalem without far-reaching repercussions in the various communities, and our studies would be carried out in a manner that would assure that the voice of each of these would be heard.*

Of course, in Jerusalem, letting every voice be heard is a full-time job in itself.

* * *

As I became more involved in academia, I also became more involved with the International Design Conference in Aspen, Colorado, an annual event that attracted a wide range of creative professionals and had considerable impact in the field. I had first attended the conference in 1967, just as Habitat opened, and given a joint presentation with Chris Alexander and Charles Correa—all three of us at the start of our professional lives. A decade later, I joined the IDCA board, becoming friends with people like Ben and Jane Thompson, who had transformed the Boston waterfront (and the waterfronts of other cities); the graphic designers Milton Glaser and Ivan Chermayeff; the designer and filmmaker Saul Bass; the industrial designer George Nelson; and Richard Wurman, who later created the TED conferences.

Meanwhile, the environment was changing dramatically in the world of architecture—a development one couldn't miss when teaching at Harvard. Many of the changes in the dominant sensibility of architecture in the 1970s and '80s can be traced back to Robert Venturi's 1966 book *Complexity and Contradiction in Architecture*, which advocated a looser, more permissive approach that emphasized the more formal, sculptural, even decorative aspects of architecture. These ideas and those of kindred spirits would eventually be given the label "postmodernism." Venturi disavowed postmodernism yet at the same time helped to sustain it: "Architects can no longer afford to be intimidated by the puritanically moral language of orthodox modern architecture," he wrote. "I like elements which are hybrid rather than 'pure,' compromising rather than 'clear,' distorted rather than 'straightforward.'" He went on to argue for "messy vitality over obvious unity" and "richness of meaning rather than clarity of meaning." As Venturi's words unintentionally demonstrate, trying to describe postmodernism is not an easy task. What was once a movement, or maybe a school of thought, now appears to have been a passing distraction. But it was consequential.

Postmodernism emerged out of disappointment with much of modernism's actual record in the postwar years— what it did and didn't achieve—but was also swayed by the influence of the world of fashion and prevailing attitudes in the world of art. The changes had been brewing for some time. In the visual arts, Pop art and particularly Conceptual art had redefined the criteria for the appreciation, even the validation, of what constitutes art. And the world of fashion has always been intrinsically linked to consumption and commercialism.

When I was starting out, the social mission of architecture—as its fundamental framework—was unquestioned. There were of course different interpretations of the mission, different readings as to how to reach the goal—I don't want to underplay the differences. Architects can get on a high horse and swing their scimitars as capably as any professional. Freud's famous phrase about "the narcissism of small differences" applies to some of these debates. But the core values seemed to be universal. I cannot imagine anybody in the 1950s walking through a school of architecture and saying, "Do whatever you want," or "It's not our priority to care about public housing," or "Architecture is just for those who can afford it."

Whereas modernists were inspired by social ideals, the approach of postmodernists was essentially pictorial. Buildings were shaped, clad, decorated with, and inspired by historical and ironic references, from classical pediments to Pop art. They might be considered in-jokes for those sophisticated enough to be in the know. A building's program, its physical nature, even its situational context were secondary.

In certain ways, the debate between modernists and postmodernists parallels the eternal tension between style and fashion, a tension that exists in most areas of life. Consider the Indian sari and the Indian tunic. They come in a great variety of fabrics and patterns, and over the centuries have certainly evolved, but as a form of dress they also have a certain constancy: style. I like to wear collarless white shirts made of Egyptian cotton of varying weaves. They are cool or warm when I need them to be, and they can be worn casually or formally—buttoned up with a suit. They are functional, adaptable, and durable, and possess a simple beauty. They are grounded in certain needs and values. Again: style.

Fashion is different—edgy, ironic, ever-changing, and striving to shock in an already-saturated media world. It can be striking and technically sophisticated. On occasion it has lasting significance. But mostly it is untethered, momentary, and never at peace. It demands and perpetuates continual consumption. And it becomes obsolete with the next stiff breeze.

Architecture's abandonment of a social mission made it all too easy for the postmodernists to cut a swath through the profession. Modernists may not always have achieved the results they desired, but for a modernist, urbanism and architecture were inseparable. One begat the other. If one wanted to promote a particular high-rise-building typology, then one really had to think about the urban environment this high-rise building would inhabit and produce. That the modernists often failed in their actual realizations—the "towers in parks" idea gave us generations of failed housing projects devoid of meaningful urban life—did not mean that the aspiration to address society's greatest needs should be abandoned.

The postmodernists would have none of this. They claimed to be unable to solve society's problems and urged an embrace of the urbanism of the past, which, at least in their view, "worked." In his own slightly dotty way, a leading voice for this view was the Prince of Wales, urging a return to what made nineteenth-century cities great. But nineteenth-century cities didn't have automobiles. They didn't have to deal with the high-rise as a building type. They also had poor standards (for instance, with regard to sanitation) and conventions about style that people today would not accept. A nostalgic approach to architecture could not accommodate the profound changes that had occurred in the nature of urban life. Postmodernism just gave up on trying to confront these difficult issues. That's why many architects opted out of dealing with housing, which happens to be one of our most basic needs. Housing as a focus for architecture has lately experienced a resurgence of interest, but it was a dormant subject for decades. In the meantime, what took the place of a social mission was an attitude of permissiveness. It was as if architects were saying, if we can't change the world, then let's at least do what we are good at: design. In time, Philip Johnson became the chief propagandist for the movement. He once observed: "There is only one absolute today and that is change. There are no rules, surely no certainties, in any of the arts. There is only a wonderful freedom."

As this took hold, I began to feel increasingly isolated at the Graduate School of Design, where postmodernism had become influential if not dominant. I came to architecture in a period deeply shaped by modernism, but as architecture has evolved, I don't find myself fitting easily into any one category—and do find myself alert to the limitations of labels and "isms." In frustration, I made use of other outlets. In 1979 I organized a debate about postmodernism at the Aspen design conference. I had been elected chair of the conference that year, which gave me the opportunity to choose the theme and to plan the program. I decided that the debate should not involve only architects on both sides of the fence but also people in other disciplines. I invited, on the architecture side, people like Robert A. M. Stern, who was a vocal proponent of postmodernism, but also Chris Alexander and others who believed that architecture must support deep human needs and have a humane social ethic. Also invited were fashion designers, such as Issey Miyake, who was then a rising star in Japan and whose clothing seemed to reflect a deeper mission, but also Pauline Trigère, a French haute couture designer who for her presentation took a piece of fabric, cut it up on stage, and pinned it into a dress—an amazing performance. Stephen Jay Gould talked about the evolution of design in nature, and Paul MacCready, who had designed the Gossamer Albatross, the lightest aircraft that you could fly using human power only—it had recently crossed the English Channel—also gave a presentation. The mathematician Benoit Mandelbrot discussed his recent invention of the science of fractals, the mathematics that explains the complexity of surfaces and lines (such as a coastal edge) in nature. Fractals proved to be useful to the world of design.

In my own investigations, the breaking up of the surfaces and volumes of a building is what we were exploring in the Habitat projects.

The Aspen gathering was an extraordinary assemblage. The architecture critic of the *Washington Post*, Wolf von Eckardt, who attended the conference, spent some time rhapsodizing about the Gossamer Albatross before getting to the point.

> *Everyone seemed agreed that orthodox, abstract Modern, a.k.a.*
> *The International Style, has failed to attain its ambitious*
> *social goals and that it sinned by divorcing itself from history.*
> *The issue was only whether history can be treated as a kind*
> *of decorative adhesive, tacked on otherwise sterile buildings.*
> *That, in essence, is what the Post-Modernists, a small but vocal*
> *group of young architects based mostly on Manhattan, seem*
> *to be doing, as architecture students around the country and*
> *architecture critics in New York watch with fascination.*

The title of the conference was Form and Purpose, which became the title of the catalogue and also of a book I would soon publish. As I wrote in the book:

> *One could not deal with form and purpose without*
> *investigating the relationship of form to purpose in nature;*
> *without reviewing the history of their relationship*
> *in manmade design throughout the centuries; without*
> *exploring the basic human urge to decorate and to celebrate*
> *the phenomenon of ritual; and last but not least, without*
> *confronting the omnipresent force of fashionability, the striving*
> *for novelty and constant change.*

As the debate in the profession heated up, I decided to publish an article about postmodernism in a venue that would get attention beyond a limited architectural audience. William Whitworth, the editor of *The Atlantic*, invited me to give it a try. Rather than simply level criticism in general terms, I thought that giving specific examples of the kind of work I found fault with would make my objections clear. This invariably led to having to name names: Charles Moore, Stanley Tigerman, Aldo Rossi, Robert Venturi, and others. As I noted in the article, drawing from historical precedent was not the issue—Le Corbusier was inspired by the ancient Greeks and the Mediterranean vernacular, Louis Kahn by medieval Europe, Frank Lloyd Wright by traditional Japanese architecture. The problem arose when history, or references to historic styles, became nothing more than a mix-and-match fashion accessory, an anecdotal reference—private jokes for those clever enough to get them.

The impact on students was profound; it was seductive to treat history as a bag of tricks. Rather than understand the causal as well as historical evolution of building and urbanism and the relation between context and form, [students are] allowed, indeed, encouraged, to dive in and pick a motif out of history to be reproduced and elaborated. Thus we find parts or wholes of buildings—or sometimes even details, such as moldings— reproduced and transformed in scale, not only out of context but without the understanding of what brought them about in the first place.

A 1980 book distilled my beliefs—including about the wrong turn I believed architecture had taken.

I have been close to Frank Gehry since the 1970s, and he was one of the people I critiqued in the article. Years later, over dinner in Los Angeles, he said to me, "You have no idea how much anger you aroused in that piece. People were just not used to being told off by colleagues in public, and certainly not attacked in such a strong, direct way. I was also upset."

The brilliant title given to the article by the editor of the magazine—"Private Jokes in Public Places"—didn't help unruffle any feathers. But the issue that was involved—which goes to the very purpose of architecture—was fundamental. It was about who we are and what we do as a profession. I returned to the debate again after the completion in 1984 of Philip Johnson's AT&T Building in Manhattan, with its Chippendale top and other dilettantish appropriations— writing him a letter in which I criticized the building's lack of a "deep sense of commitment." I went on:

It is, however, the total attitude behind this building which I would like to challenge. . . . If I were to attempt to summarize what I consider this attitude to be, I would say it is a preoccupation with the arbitrary, visual aspects of a building as the main generative consideration in its formation. This formalistic (and sometimes eclectic) visual game in which there are no rules is motivated sometimes by whim, humor, or a desire to shock, and sometimes by boredom or nostalgia. It appears not to be based on any consideration of lifestyle, let alone on the materials, processes, and physical realities which underline the raison d'être of the building.

Johnson wrote back, blithely unyielding:

Eclecticism *is not a bad word to me.* Visual aspects *of a building seem, to me, primary (are my arches more arbitrary than your arches in Jerusalem?). In counter distinction to the theoreticians of the heroic period, it seems to me that* form *is everything. The very raison d'être of a building seems to be* form.

This was not a meeting of the minds. There was a time when Johnson used to invite me to the small gatherings he convened at his home in Manhattan to talk about architecture. Those invitations stopped. Johnson's opinions had considerable influence, for a period of time. He had a lot to say when it came to the selection of architects to serve on commissions and juries or to become the subjects of museum exhibitions.

Some years later, my son, Oren used *Private Jokes, Public Places* as the title for a play about sexism in architecture education. (Oren had pursued architecture before becoming a playwright.) The play had a successful run in New York and has frequently been revived. Students at architecture schools sometimes perform it as a way of starting the academic year. Postmodernism may be an "ism" that has faded away, but the permissiveness that arrived with it prevails to this day.

* * *

Only after I moved to Boston did I begin to develop relationships with the commercial world of real-estate development, as opposed to pursuing projects where the client was a government or an institution or, in a few cases, an individual. My introduction to that world came through Mortimer B. Zuckerman and his company Boston Properties. I had met Mort at McGill, where we both were students.

When I came to Boston, he was one of a handful of people I already knew. Mort is charming and very smart—a real intellect—and also aggressively ambitious. He wanted to be not only a major real-estate developer, as he was, but also a cultural force, which is why he bought *The Atlantic* and, later, *US News & World Report*.

Mort lived at the time in a brick town house on Beacon Hill. He employed a celebrated Chinese chef and presided over a weekly evening salon—writers, journalists, scholars, artists, politicians. Mort introduced me to Boston's mayor, Kevin White, which led to my involvement in a number of proposed local ventures. Through Mort, Michal and I met Marty Peretz, who was the owner and publisher of *The New Republic* and, at the time, an intellectual force at Harvard; in 1981, Michal and I would be married in the garden at the home of Marty and his wife, Anne. Before long, our daughter Carmelle was born, followed a year and a half later by our daughter Yasmin. Now we were a family with young children in our lives—a major change.

With Mort, our personal friendship, which became very close, soon acquired the added dimension of a professional relationship. Mort was then involved in two major Boston-area projects. One involved parcels of land abutting the historic Boston Public Garden—a controversial project for a mixed-use residential/hotel complex. It never won approval. The other was Cambridge Center—an office, retail, and hotel development in Kendall Square, right across the Longfellow Bridge from downtown Boston, adjacent to MIT—which had stalled under the skeptical eye of Cambridge planning authorities. At Mort's direction, I took over the Cambridge project from a New York firm, Davis and Brody, and came up with a new scheme. It probably helped that I was the new boy in town—it takes a while for the initial cloud of goodwill to evaporate.

But I also took a different approach: whereas the Davis and Brody plan placed a large parking garage along Broadway—destined to be a lively pedestrian street—my plan buried the garage in the center block and surrounded it with buildings that animated the sidewalks all around. And, as a bonus, we proposed that the roof of the parking garage, four floors above street level, become a public park for the thousands of office workers in the district.

With the developer
Mortimer B.
Zuckerman,
1985.

The plans were approved, and we built several office
buildings, a Marriott hotel, and the Harvard/MIT Coop,
along with a subway station and a piazza. Kendall Square
was developed under very tight constraints, to the point that
there was very little room for architectural adventure. It's
good—it's solid urban design. Go to Kendall Square today,
and it's bustling with activity and life, fully built up, the trees
mature—a vibrant urban district, our own work at its core. It
is a success, but the word "transcendent" is not one I'd apply.

Another developer with whom we worked was Richard
Cohen, then based in Boston, who owned land in Cambridge
facing the Charles River. He wanted to build a residential
complex, to be called the Esplanade. Here was an opportunity, right on the river, to design a mini-Habitat in a setting
somewhat similar to the original. We terraced the twelve-
story brick building, creating private gardens that cascaded
toward the river with spectacular views of the Boston skyline
and the Massachusetts State House with its golden dome. A
row of town houses, also facing the river, concealed two levels
of parking, so that the garage, even though aboveground,
was invisible. The roof of the garage became a communal
garden. In a brilliant marketing move, Cohen sold one of
the first apartments to the president of Harvard, Derek
Bok. In an academic setting like Cambridge, that might as
well have been a neon advertising sign.

Meanwhile, my association with Mort Zuckerman suddenly brought New York City into view. Manhattan had not
been a central focus of my work. I had explored some Habitat
possibilities in the late 1960s, when John Lindsay was mayor,
but the plans had remained on paper. A few years later, I
teamed up with the developer Samuel LeFrak to conceive
a design for Battery Park City, a virgin landfill site in lower
Manhattan. The project was eventually smothered by that
city's familiar combination of opposition, bureaucracy, and
the intrusion of other interested parties, with or without a
shiv. Even so, I will never forget the unveiling of the model
on Sam LeFrak's yacht as we cruised the Hudson at sunset,
accompanied by financiers and journalists.

In the mid-1980s, Boston Properties began shifting its
focus of activity away from Boston and into New York and
Washington, D.C. It would eventually move its head office,
though the company would keep its name. In 1985, Mort asked

me to join him in the competition for developing Columbus
Center, facing Columbus Circle and Central Park, in New
York: three million square feet of offices, hotels, residences,
and retail, straddling a hub of subway lines on a site long
occupied by a crumbling convention center, the New York
Coliseum. New York's Metropolitan Transportation Authority
(MTA) was the owner of the property and was putting it out
for proposals. Everybody who was anybody in the development business wanted Columbus Circle, and all the major
players had paired up with architects.

I came in with Mort, who joined forces with Salomon
Brothers as the prime tenant, a real coup, given that the
investment firm was, at the time, the "king of Wall Street."
The decision by the MTA would be based on both money—
how much the developer was willing to put up for the land,
which in Mort's case would be $455 million—as well as de-
sign. We offered a number of compelling design strategies.
Perhaps the most significant move was a decision to leave
the Fifty-Ninth Street view corridor open, west to east, by

A notebook sketch
for the Columbus
Center project,
1985.

placing two towers on the site (rather than one, as others did), framing the view and allowing sunrise and sunset to be experienced along Fifty-Ninth. In deference to Columbus Circle, we decided to strongly define the actual circle by placing a five-story curved gallery along its circumference, with shops facing the circle. Crowning this galleria was a four-story triangular glazed atrium accommodating the Salomon Brothers' trading floors.

The towers were to be fifty-eight and sixty-eight stories high. One was dedicated to Salomon Brothers, the other to the Intercontinental Hotel. The towers bridged at level thirty-nine, above which they were both residential. Portions of the towers stepped-back every ten floors, each setback topped by turret-like four-story glass enclosures containing gardens to serve the buildings' occupants. The residential units themselves were terraced, providing Habitat-like gardens and creating a hillside village in the sky.

A formal model of Columbus Center, at the southwestern edge of Central Park, 1985.

It was an exciting project—almost a dream. And although Boston Properties did not offer the highest monetary bid, its design was selected as the winner. The total cost was estimated to be $1 billion. I opened an office in New York City and hired a team of forty people to work on this project alone. The teamwork required when undertaking a major architectural effort is immense. The project is led by the design leader—my role— who is part composer, part conductor, part movie director. Key players include the structural engineer and the mechanical engineers (who are in charge of the design of the building systems). Landscape architects, lighting designers, graphics-and-signage designers, and acousticians are all part of the mix. The actual work of building involves contractors, specialist subcontractors, superintendents, craftsmen, and thousands of laborers. All these components must be able to work together, making continual adjustments and solving unexpected problems under conditions of maximum stress—while at the same time keeping in mind the needs and sensitivities of the client.

The mood in our New York office was reminiscent of the Habitat '67 days, devoted exclu-sively to one project. All the other projects were being done out of Bos-ton. I shuttled between the two cities. A vast meeting room was a place of endless consultations with armies of consultants and would-be tenants. Not long ago, I had reason to review some of the sketches and design stud-ies from that time, and the wealth of material was staggering. Columbus Circle clearly consumed me.

Reflecting on this period, I am struck by how professional and personal trauma came together with brutal force. In 1987, as work went forward on Colum-bus Center, Michal was experiencing a difficult pregnancy. We had two daughters—Carmelle, then five years old, and Yasmin, then three—and an early amniocentesis revealed that Michal was now carrying a boy. We later learned that there were serious medical issues. Though anxious, we continued with the pregnancy. Eventually it was determined that the baby was too malformed to survive. Given the lateness of the diagnosis, it was difficult to find a doctor in the United States willing to terminate the pregnancy; the procedure was performed at the Hadassah Mount Scopus Hospital in Israel. Both Michal and I were profoundly burdened by the decision we took, though we never questioned it. We carried forward a deep pain and sense of loss.

Cartoonists weighed in as controversy mounted. In the pages of *The New Yorker*, King Kong wondered about finding a foothold.

That was the wrenching personal backdrop as the Columbus Center project began to run into staunch oppo-sition. The project was of course very much in the public eye, and our design was getting clobbered by outspoken critics. It began with a blast by Paul Goldberger, the architecture critic of the *New York Times*. Others piled on. There was no mistaking where Goldberger stood. As he wrote at one point:

> *Mr. Safdie's building is slice-and-dice architecture. The design would have a series of slanted roofs which would meld into a gangling composition of anxious angles, inappropriate for the*

corner of Central Park not only in size, but also in form, mass and detail.

Goldberger was in his postmodern period, and I think he wanted the building to somehow echo the turreted twin-tower motif that runs along Central Park West; a kind of pseudo-contextualism that a postmodernist might pursue. For Salomon Brothers, criticism was like water running off a bird's feathers—they could not have cared less. But criticism was not what Mort Zuckerman wanted to hear. Like most people, and perhaps more than most, he preferred to be loved. We were both invited to an editorial board meeting, one of those famous lunches, at *The Times*. We made our case. Afterward, as we walked along Forty-Third Street, Mort said, "Moshe, why can't you redesign to make them happy? Why can't you do something that is more in the spirit of what they're after?" He was feeling the heat. I told him, "You know, not only can't I do that—it's not the way I am—but I don't think you could design to order even if we wanted to. Because whatever we do, does not mean that they're going to endorse it."

The battle was soon joined by the Municipal Art Society. The society took aim less at the design and more at the project's mass and height. It managed to recruit Henry Kissinger and Jacqueline Onassis and other notables as the public face of opposition. They led a parade of dignitaries, all holding unfurled black umbrellas, through Central Park to protest what they described as the shadows that the buildings would cast over the park at certain times of day. I will admit that the Municipal Art Society's performance-art strategy was brilliant. So was making the argument about mass and height, which was at least more objective than any argument about design. As a practical matter, there was little to be done about scale. The plot of land was a certain size, and a certain number of square feet had to be built, consistent with zoning, if the building was to be profitable. To this day I can't fully account for the resistance to the Columbus Center plan. Some of it may truly have been aesthetic, perhaps reflecting architectural tastes of the moment. Some of it may have had to do with Mort, and perhaps myself, both being outsiders in New York. And some of it certainly was a worry about mass and height.

In the end, though, what dashed our hopes was not the ongoing criticism and controversy but the Wall Street crash of October 1987. The project was still going ahead full steam. We had finished the working drawings, had ordered steel, and had selected stone. And then the bottom fell out of the market. There was panic at Salomon, which abruptly and dramatically withdrew from the lease, a decision that would ultimately cost it $50 million in cancellation charges.

With Salomon gone, Mort recognized an opportunity. I was in New Haven, Connecticut, at a design conference at Yale, when I got a phone call. The call was not from Mort but from his partner, Edward Linde, who said, "Moshe, this is it." I said, "You mean you want me to do some redesigning?" "No, no," Ed replied. "This is it—we're going to change architects." I was in shock. Someone in our office recently found my sketchbooks from that month. The day of the call was Friday, December 4, 1987, and on those pages in the sketchbook are stylized geometric designs, all seemingly on fire. The sketches are labeled "Black Friday."

Though it did not affect my functioning, I know that, in the aftermath, I experienced something akin to depression. This was the most visible project in New York City at the time, in the heart of Manhattan. I had to lay off forty people in one day. And it hurt that Mort, a longtime friend, had not broken the news to me himself. Mort hired David Childs of SOM to take over the project. Childs was also president of the Municipal Art Society. And he was a gentleman. He would not start work until Mort paid our firm what we were owed.

As time went on, Mort would lose the project altogether. What was first known as the Time Warner Center, and then as the Deutsche Bank Center, now stands on the Columbus Circle site. It still has two towers, but they are taller than mine would have been, and therefore cast bigger shadows. Meanwhile, a number of supertall apartment buildings have been allowed to rise on the south side of Central Park, casting northerly shadows that fall even more directly on the park. Where was the Municipal Art Society while all these projects were being approved and constructed?

I did not see Mort Zuckerman again for years. I once thought we'd had the sort of friendship that could handle tough moments, but when a tough moment came, he was not up to it. I suspect he felt bad about what he had done. Our mutual friend Marty Peretz reported that he had scolded Mort for possibly destroying my career. Mort responded, "Moshe will be OK. He's about to open the National Gallery of Canada. He'll recover." The next time I encountered Mort—the only other time—was in Israel. Michal and I were attending a dinner in honor of President Bill Clinton at the residence of Yitzhak and Leah Rabin, the prime minister and his wife, who had become good friends. When Clinton and his party arrived, Mort was with them. I remember the meeting as being marked by the forced cordiality that deep awkwardness can produce.

* * *

One of the great bonuses of the move to Cambridge, and to Harvard, was that a new and impressive social and intellectual community became accessible. Within the Harvard community, the business school held a special place. It was certainly the richest school in a university where each faculty and school was a "tub on its own bottom." It was the only major program, with the exception of athletics, that lay across the Charles River, on the Boston side. The business-school campus had been planned and designed in 1928, on the eve of the Great Depression, by the famous New York firm of McKim, Mead & White, the architects of Manhattan's original Pennsylvania Station and Town Hall theater, and the Boston Public Library. The plan was typically Georgian, with the individual buildings forming courtyards and thereby creating a cohesive urban fabric. Brick with stone trim, the buildings are restrained but elegant. I often reflect in amazement at how good one feels just strolling through the business school's campus—the scale, the interface between the individual buildings, and the overall layout are compelling. It all works.

In 1979, a year after my arrival in Boston, I received a call from John H. McArthur, the dean of the business school. We met at his office and instantly connected. It helped that we shared a Canadian background: McArthur, a onetime forestry major and an alumnus of the business school, was originally from Vancouver. He laid out his concerns. Activities had intensified on the business-school campus, causing the buildings to be adapted willy-nilly to meet the needs

of modern life. Many elements of the campus had begun
to deteriorate. Each individual building had its own small
exterior air-conditioning plant, its own loading dock. Vehicles
traveled freely on paths conceived for pedestrians. Meanwhile, the faculty had become too dispersed, the members
sequestered in their own mini-fiefdoms. There was not enough
dialogue, not enough interaction.

Finally, McArthur came to an issue that lay somewhat
outside grand strategic planning but that intrigued him greatly.
The original McKim, Mead & White master plan had made
provision for a chapel, but it had never been built. The site
was vacant, and he felt it was time to remedy that oversight.

My team and I took on the full range of planning responsibilities. Among other things, we returned the campus to its
pedestrians-only condition. We consolidated the mechanical
systems. We connected the buildings by means of a network
of underground tunnels—not for pedestrians but for deliv-
eries and mechanical services. We also expanded L-shaped
Morgan Hall by adding a second L, forming an atrium courtyard between the two. This provided sufficient area to create
individual offices for all faculty in one central structure. In
the center of the atrium, we placed a fourth-century Roman
mosaic that had been at Dumbarton Oaks, in Washington,
D.C., illuminating it with computerized pivoting mirrors
that directed sunlight onto the tiles at all hours of the day.

The great challenge was the chapel. McArthur did not
want to burden Harvard's usual fund-raising channels, so
he contacted his classmates—the class of 1959—who came
up with the money. I was given no program, except that
the chapel would be nondenominational, but I began to
develop a program in my own head. The chapel would serve
the students at large and it would be a place of worship and
memorial, serving any group that might be on the campus. Because the site was at the gateway to the business
school, it enjoyed a powerful and symbolic presence. The
idea of a chapel at the wealthy business school occasioned
more than a few jokes. Was the chapel about the worship
of God or of mammon?

We thought that some students would want to come,
relax, read a book, and meditate without necessarily entering
the inner sanctum itself. We therefore proposed a glazed
garden, delicately arranged with biblical plants and scented

flowers; the garden would step down into the earth, enclosed by a glass pyramid, its lowest level holding a fishpond. The soothing, uplifting effect of water on one's mood is surely primeval. So is the power of scent, often ignored in architecture but in this instance imagined as a fragrant doorway to what lay beyond, lush with olive trees, jasmine, wisteria, climbing roses.

The garden led to the chapel. It would be an introverted space—circular in plan and sculpted with twelve concave apses. Because the apses surrounded the perimeter, services or events could be organized flexibly—to face Jerusalem for a synagogue, to face Mecca for a mosque, or to face any direction at all for a Christian service. You could sit in the round or all look one way, toward a pulpit or an altar.

The prism effect: the Class of 1959 Chapel at the Harvard Business School, Cambridge, Massachusetts, completed in 1992.

I wanted the space to be sublime and yet also dynamic—somehow changing in the course of the day, even though there were no windows giving views to the outside. The chapel would be lit by a series of skylights, each set above an apse.

I recalled a proposal I had made some ten years earlier to the rabbis of the yeshiva in Jerusalem. They had asked me at the time why I did not incorporate stained-glass windows into the synagogue. They had seen such windows in an earlier rendering that had come from a New York architect. I responded that stained glass was really part of a Christian tradition and relied on storytelling though religious images, which was frowned upon in the Jewish tradition. They dismissed my reaction and got around the objection by suggesting that the windows have an abstract design.

I thought about that episode in the context of the chapel. What could be more abstract than light coming through giant prisms set in the skylights, which would break white sunlight into the full spectrum of colors? Because the sun is continually moving in the sky, the patterns of colorful lights would constantly shift on the walls and the floors. I remembered reading about an artist in New York, Charles Ross, who was making very large prisms, and he turned out to be interested in the project. Ross made prisms for the chapel out of acrylic sheets—triangles some two feet across and ten feet long, the void filled with mineral oil. They worked beautifully. The end result was a cylindrical building topped with prisms and entered through the four-sided glass pyramid that enclosed the garden. For the cladding we used copper. It takes a decade or more, depending on air quality, for copper to oxidize and attain that rich, turquoise-green patina that is so deeply attractive. McArthur was impatient. We found a Swiss company that could pre-oxidize the copper sheeting, with the result that the chapel looked the way it would "become" from the moment it was built.

We had never planned specifically for the chapel's acoustic nature, because the space is relatively small, and we anticipated that it would be used only for speaking and the kind of music that accompanies a service. But something about the concrete walls and the overall geometry turned out to provide superior acoustics, and concerts are now held there in addition to religious services.

I will never forget the moment when I first understood the chapel's fine acoustics. As we were finishing construction, I brought over my friend Yo-Yo Ma to see the chapel. He had his cello with him, as he was returning from rehearsal. We walked through the garden into the chapel and sat on the simple Amish-style chairs that had been selected. It was a sunny day, and a spectrum of colors was dancing on the walls. Yo-Yo opened his case, pulled out his cello, and started playing Bach's Cello Suite No. 5 in C Minor. After a few chords, listening, he paused and said, "This is a miracle!"

Looking skyward from within the Great Hall at the National Gallery of Canada, in Ottawa, Ontario, completed in 1988.

CHAPTER SIX

"Does God Live There?"

An uplifting of the spirit is something I have always aspired to in my work. It has taken form in small ways—private ways—as when I redesigned my own house in Cambridge, and in somewhat larger and more public ways, as with the Class of 1959 Chapel at the Harvard Business School. And it can be on a grand scale, as I hoped to achieve with the National Gallery of Canada, set on a riverside promontory in Ottawa.

The path to the National Gallery was not a straightforward one—there were miles of twists and turns. I had been active in the United States, Israel, and elsewhere for more than a decade, but ever since Habitat '67, Canada had for some reason been tough to crack. To this day, I can't quite explain why this should have been so. To be sure, the advent of the separatist movement in Quebec created a difficult environment for non-French Canadians. Life in Montreal was changing. There were separatism-fueled referenda, one after the other. Banks and other big companies started leaving. Jewish professionals, particularly doctors, also began moving away. I inhabited some category even further into the wilderness than persona non grata when it came to Montreal's mayor, Jean Drapeau.

I could understand why I was not getting work in Quebec, but the indifference of the national government was more of a mystery. Nor was the lack of access to the business community easy to understand. But there was no work to be had. I was aware of a certain vibe, which may have been real or may have existed mainly in my head, that I was seen as a wild young dreamer who had done that crazy Habitat project back in the 1960s, and who was not suitable for grown-up engagements.

In 1976 came the announcement of a competition for a new National Gallery of Canada, to be built in Ottawa, on a site next to the Supreme Court, overlooking the Ottawa River. Many architects in Canada applied to be included in the competition, and the sponsors produced a short list of twelve architects to go forward with designs. I was not included among the twelve, which surprised and troubled me. This was three years after I had traveled to China with Pierre Trudeau, the prime minister. Trudeau was an art lover and deeply appreciative of architecture. As I came to know him personally, I also came to see him as a statesman rather than simply a politician. I wrote him a letter—

perhaps a little more emotional than usual—saying that the situation seemed strange. "Is it not ironical," I wrote, "that the National Gallery would have found it appropriate to hold an extensive exhibition of our work through the summer of 1974 (the only one of its kind of the works of a Canadian architect, to my knowledge), and yet when it comes to constructing and designing the building for the National Gallery, we should be excluded?" He wrote back saying that it was indeed strange, and he invited me to meet with the minister of public works to explore why I seemed to have been shut out of government commissions. Nothing came of that meeting. In the end, the National Gallery competition was won by John Parkin, whom I knew from the Habitat '67 experience; and then, unexpectedly, the project was shelved.

Jean Sutherland Boggs was the director of the gallery at the time—the first woman to hold that post and a transformative leader and scholar. Her specialty was Edgar Degas, and she wrote about him with insight and rigor, but she also had a knack for public outreach. In the 1960s, she had hatched a brilliant scheme to engage the public: if listeners to the Canadian Broadcasting Corporation's radio programs would send her one dollar, she would send back a packet of ten postcards with pictures of famous works of art—and then broadcast radio lectures keyed to each work. Boggs had seen a new National Gallery as another way to connect with the public—and also, architecturally, as an enhancement of the public realm. She was deeply disappointed when the project collapsed and worn out by all the red tape and second-guessing. She resigned her job as director and went to Harvard, teaching there for several years. Then she was named the director of the Philadelphia Museum of Art—the first woman to hold that job too. As for me: I had left Canada by then and was devoting myself to projects in Israel and the United States, as well as plans for projects in Iran and Senegal.

Then, in 1981, out of the blue, I got a phone call from Claude Belzile, of Belzile, Brassard, Gallienne, Lavoie Architects. A design competition had been announced for the new Musée de la Civilisation, to be located in Quebec City—a museum that would necessarily place an emphasis on the province's own identity and history, a centerpiece ambition of the government of Quebec. Belzile and his firm wanted me to join them in the competition. It was a challenging project, and although I was skeptical, given my previous experiences in Quebec, we signed on as the lead designers. There was one momentary distraction, perhaps symptomatic of the residual thinking in some quarters behind the scenes. We had no sooner submitted our joint application than a number of Quebec architects petitioned that I should be disqualified because I was not living in Quebec. The complaint did not get traction. I was a member in good standing of the Ordre des Architectes du Québec, and I had retained my Canadian citizenship. Ultimately, we won the competition, and I started commuting from Boston to Quebec City.

The site of the new museum was an important one—it lay in Old Quebec, at the bottom of the escarpment and right on the Saint Lawrence River waterfront. The idea of incorporating a grand staircase of some kind had been in my mind from the beginning. When we researched the site's history, we discovered that it had been the location of the old market, a classical building with a pedimented portico and columns. It had featured a grand staircase in front that descended all the way into the water, so that products could be unloaded from boats and carried directly inside. Farther back, rising some 150 feet, was the escarpment and plateau of the walled Upper Town. From these heights the Château Frontenac hotel overlooked the Saint Lawrence and the Lower Town with its old monasteries and churches. The Lower Town was medieval in scale, with tight narrow streets and a small harbor.

I had explored the concept of a grand staircase before— in the design for the student union at San Francisco State College and in the design for the Centre Pompidou. In each case, you could both climb upon the building as well as enter it at street level. As we conceived it, the Quebec museum was essentially a four-story rectangular building that filled an entire city block, out of which we carved a series of stairs and stepped gardens that connected the roof and the river. Going up the main staircase would create the sensation of ascending toward the upper city. In the competition documents, we had been informed that there were two historic buildings on the site and that both would need to be preserved and incorporated. One was a nineteenth-century stone house. The other was a hundred-year-old classical bank building. As we started construction, we also discovered the remains of the city's ancient river wall. We excavated the remains and incorporated them into the design of the museum. In important ways, building on this site offered a powerful reprise of my experience in Jerusalem: it presented the challenge of constructing something very new in an environment that was very old, and also the challenge of preserving and incorporating important archaeological features. I did not see these as constraints; for me, the impulse was the opposite: I almost wanted to dance with them.

Much centered on the skyline; from the roof, there was a magnificent view of the river and the shore on the other side. Importantly, the roof would be seen from the Upper Town, so its design was crucial. A series of ridges crowned with skylights broke the scale so that, when seen from above, it resonated with the roofs of the smaller buildings all around. The ridges were designed to facilitate the shedding of snow in this wintry city. The walls of the building were clad in limestone and the roofs with copper; a lookout spire related to the city's many church spires and bell towers. Once inside the building, large exhibit spaces were bathed in beautiful soft light—and then, surprise, visitors came upon the two old buildings, effectively encountering a story within a story.

The museum opened in 1988 and was an immediate success. It represented an important step for me back into Canada. I was surprised at how strongly I had identified with the objective of creating a building that was quintessentially Quebecois.

We were still mid-construction when another opportunity arose. Prime Minister Trudeau, no doubt thinking about his legacy, decided that he would build two new museums

The old market and the staircase descending to the Saint Lawrence River in Quebec's Lower Town, 1874.

Roughly the same view, with the Quebec Musée de la Civilisation—and staircase—on the site, 1981.

in Ottawa. One would be a history museum—the Canadian Museum of Civilization (initially "of Man," and now "of History"), which would be located in Hull, on the Quebec side of the Ottawa River, across from the Parliament of Canada. The second, reviving an old dream, would be the National Gallery, which would be positioned on a high, prominent site across the river from the Canadian Museum of Civilization, adjacent to the Notre-Dame Cathedral and facing Parliament, opposite a deep river inlet. To show he meant business, Trudeau lured Jean Boggs back to Canada and appointed her chair of the newly established Canadian Museum Construction Corporation, the entity that would select the architects, be responsible for construction, manage the budgets, and in general oversee the entire enterprise. Boggs launched the selection process, and did so noting that selection would be based largely on an architect's overall work and sensibility rather than on any specific proposed design—though we would need to submit one. Ultimately, she and her committee came up with a short list of twelve candidates for the two projects. This time, my firm was among them. At stake was the opportunity to create not only a new type of museum building but a building of national significance.

* * *

Jean Sutherland Boggs turned out to be one of the greatest clients I have encountered. To understand what I mean requires understanding something about clients as a species. To an architect, a client can be an individual or an institutional committee; a client might be representing a government, a private institution, or simply themselves. Sometimes a client may be each of these at once. On occasion—and almost always when working for large corporations or in China—there may be parties involved who remain in the shadows, forceful yet invisible.

Over the years I have come to realize that clients can be placed into a limited number of categories. There are those who, from the initial moment, exude a feeling of total harmony, a sharing of higher vision and concrete objectives. That doesn't mean we think the same way or always come to the same conclusions. It doesn't mean that there isn't intense debate. But in a larger sense, we understand that we are after the same thing, and share values and even basic feelings about taste. This kind of client is often able to lift an architect to places he or she would not otherwise be able to reach. Alice Walton, the Walmart heiress who commissioned the Crystal Bridges Museum, in Bentonville, Arkansas, falls into this category. So does Nancy Tessman, the former director of the Salt Lake City Public Library system. And so does Jean Boggs, whose instincts on certain key decisions about the National Gallery proved to be shrewd and correct.

Another kind of client is the one who has good intentions but is not very experienced or versed in the process of building, and yet is open to learning. I don't say that in a condescending way; all of us are in this same boat with respect to aspects of professional life that are not our own. This kind of client may not have much in the way of intrinsic sensibility or tools, such as the ability to easily read plans. But he or she is willing to spend the time needed to learn, and together we arrive at the right place. Uri Herscher, the founder and longtime president of the Skirball Cultural Center, in Los Angeles, is such a client. In the course of twenty-five years of collaboration on many phases of the Skirball, we grew and educated one another to where we are today.

And then there are clients who are motivated by things other than what motivates the architect. I am thinking, for instance, of commercial clients who, whatever their design sensibilities, may have an unswerving commitment to something they see as more important: the bottom line. Commercial clients are also more inclined than others to be know-it-alls, confidently insisting that there is only one way to do things, which not surprisingly turns out to be *their* way. Sam LeFrak, the developer with whom I worked on that never-realized plan for Battery Park City, in Manhattan, knew his business, and he always spoke with New York–style candor. He possessed a good-heartedness that helped to mellow his toughness, and I liked him. He once said to me: "Moshe, remember the golden rule in design: he who has the gold makes the rules." It would be misleading, though, to conclude that clients driven by the bottom line cannot be great patrons. Liew Mun Leong, the former CEO of CapitaLand and former chairman of Changi Airport Group, enabled our firm to realize a number of landmark projects: Sky Habitat, in Singapore; Raffles City Chongqing, in China; and Jewel at Changi Airport, also in Singapore.

After six decades as an architect, I've learned how to cope with the golden rule and many other rules. My instincts are better tuned now than they were a half century ago, and I pay attention to them. On rare occasions, meeting a client sets off an alarm bell—I know immediately that the relationship will not work and that I must decline.

The interplay among the client, the mission, the finances, and the design never becomes truly second nature or something you can take for granted—every project brings a different constellation of challenges. Louis Sullivan once observed that great buildings occur only with great clients, and he also observed that clients get the architect they deserve. There's a certain tension between these observations. In that sometimes frothy space is found persuasion and education and trust as well as, on occasion, intimidation and manipulation.

In the early 1960s, when I signed on with Sandy van Ginkel to help plan the world's fair in Montreal—and then to design and build Habitat—the nature of this interplay was not familiar territory. My experience as an architect was limited. I had never run an office, much less created one from scratch. My family certainly possessed a mercantile talent, and I likely inherited some of those skills. I was schooled in them when I worked with my father. But the skills, such as they were, had not been put to the test. That changed with Habitat '67. The skills were tested, and they have evolved and matured since then. To deal productively with people as culturally different as Israelis, Senegalese,

The National
Gallery's
redoubtable Jean
Boggs, 1976, with
Bernini's bust of
Pope Urban VIII.

Chinese, Canadians, Iranians, and New Yorkers—each bringing different modes of discussion and decision-making—is an education in diversity as well as in humility.

Jean Boggs was the kind of client who propels one to new heights. She was a scholar and a curator. Her stout frame suggested determination, and her bright eyes and curly white hair suggested accessibility. There is a photograph of Boggs that I love, with Bernini's bust of that great patron Pope Urban VIII in the background. It is an inspired pairing. In her deceptively slow-moving way, she glided through political storms and complex technical debates, always with a knowing smile on her face. She set the bar high.

Returning to Canada in 1982, with the full backing of Trudeau, Boggs was given complete authority to build the two museums he wanted. She wrote an elaborate program for each one. She understood museums inside out. She did her background research diligently, visiting projects by all the architects whose work she had her eye on. She was opinionated—and was not afraid of putting her foot down. At the same time, she supported doing things that had not been done before and that I thought were important, such as bringing daylight to every room in which art is displayed on any level of the building.

* * *

After the first stage of the selection process, during which Boggs identified her list of twelve architects, she introduced a wrinkle: each of the candidates was asked to state a preference about which project was of greatest interest—the civilization museum or the art museum. But the stated preference was only that. Boggs and the governing museum corporation would be free to select any of the architects for either of the projects.

My own preference was to do the Museum of Civilization. I was being somewhat opportunistic: I was in the process of completing one in Quebec and had acquired a feel for the issues involved. And I thought that, as a onetime resident of Quebec, I might stand a better chance with the museum being built on the Quebec side of the river. There wasn't much time between short-list selection and final decision, so we got right to work on a scheme. The site was flat, along the Ottawa River shoreline. Upstream was a large pulp and paper mill, still active, and logs were floated down the river for processing. Across the river and high on a cliff lay Ottawa and the Parliament complex. I decided to conceive a major part of the museum as an island within the shallow part of the river. The main public hall would be a glazed space facing upstream, the river flowing past it as if the island were a ship. This idea may have sprung from some memory of Tiber Island, in Rome, which I had seen on that influential boyhood visit. In ancient times the narrow island in the Tiber had likewise been conceived as a ship with a marble prow. Also lodged in my memory were the island palaces of Lake Maggiore. Given Ottawa's latitude, we anticipated that in wintertime the water would carry chunks of ice, which would be dramatic. The hall itself was supported by multiple slender columns, akin to the trees in a forest, alluding to the logs floating outside.

The short-listed finalists awaited the outcome of the selection process. At last, I received a call from Jean Boggs. She said that everyone had been very taken with my design for the Museum of Civilization, but that the prime minister wanted to award the commission for a museum narrating the history of Canada to a native Canadian—specifically to Douglas Cardinal, a descendant of what we identify in Canada as the First Peoples, and a rising star among Canadian architects. That being so, they wanted to consider me for the National Gallery instead—and could I submit sketches by the following Monday?

Excited about the positive news, I wanted to respond affirmatively, but something stopped me. No, I couldn't, I

A 1983 rendering of the Canadian Museum of Civilization, in Ottawa—which led, unexpectedly, to the National Gallery commission.

told Boggs. Sure, I could certainly produce something superficial, but fundamentally my work was not about images. It
was derived from a deep understanding of the program and
the site, and in this case, I was not sufficiently familiar with
either. I told Boggs that she would have to proceed the way
she had said she would: making a decision on the basis of an
architect's overall work and sensibility, not on the basis of a
specific design. We left it there.

I didn't actually leave it there, however. In the back of
my mind, I continued to worry about a vague specter that
never seemed far away: the idea that some might feel I was
not "Canadian enough," whatever that might mean. So, I
decided to make a phone call to Trudeau. I told the prime
minister that the answer to the background question—how
Canadian was I?—might well be determined by their decision.
He laughed, but committed to nothing, though something
about his tone left me feeling positive. That was on a Thursday. On Monday, I was told that we had been selected to
design the National Gallery.

Given that my office was in Boston, I needed a local architect to support our efforts—ideally, an Ontario firm. People
in the museum corporation suggested Parkin. His firm had,
after all, won the 1976 competition, ten years earlier—the one
that had been shelved. I recalled my own off-putting experience with Parkin many years before, when my invitation to
his firm to join the Habitat project had led to its attempt to
take control. But the circumstances now were very different.
Parkin agreed to provide technical services—construction

documents and permits—and I welcomed his firm to the
project as the associate architect.

Jean Boggs knew everything about art museums, and she
had thought about this new one deeply. The comprehensive
program she had laid out was a three-volume document that
covered every room in what was as yet a nonexistent structure. At the same time, for all the thought she had put into
the National Gallery, she understood that other people's
thoughts also counted. She was open to a rich dialogue.

The first thing I did was to go away with my family for
a working holiday in Guadeloupe. The project demanded
contemplation and distance—and relative isolation—and I
used the time away to think deeply and to distill some of that
thinking into models. This was to be a very large museum,
with diverse collections and a large curatorial wing—as large
as the one at the Metropolitan Museum of Art, in New York.
The trouble with large museums, the Met included, is that
one often gets lost and, after a while, begins to feel worn down
by fatigue—something that never happens at the smaller Frick
Collection or Guggenheim Museum. I thought that a way to
overcome this would be to think of the building as a micro-
cosm of a town—with streets, piazzas, parks—surrounded
by gallery buildings: more like a series of smaller museums
clustered around landmark spaces to which visitors would
always return for reorientation.

I also believed that a large, repetitive, modular building, with endlessly similar gallery spaces, as the 1976 Parkin
design had proposed, would not do justice to the collection.
Baroque art, Inuit art, contemporary art—whatever the category, it would benefit from galleries designed specifically
for that collection in terms of scale, character, lighting,
and so on. In addition, I have always believed that natural
light is fundamental to enjoying a museum—both in galleries and in public spaces—but for the National Gallery we
went further, arguing that there should also be daylight in
every office and in the curatorial spaces. In broader terms,
I argued that a contemporary museum should be inviting,
transparent, and open to the city. I used the term "seductive."
And yet, because it would occupy a position of prominence
in the nation's capital, it would also have to be a place of
ritual and ceremony.

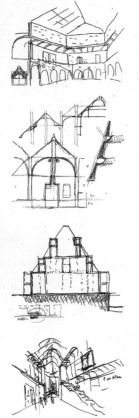

Notebook sketches from 1983: designs for the National Gallery of Canada begin to take shape.

I came back from Guadeloupe with three models that I had made myself out of Plasticine. One scheme was very much in the spirit of I. M. Pei's East Building of the National Gallery in Washington: a large central atrium surrounded by several structures, each one anchoring a key location on the site. The second scheme was more linear, consisting of two main concourses—one of them parallel to the river, one perpendicular to it—linked by a glazed Great Hall. The Great Hall would be a grand public space suitable for large gatherings of many kinds, including formal state events. The third idea was nicknamed the "village scheme"— pavilions scattered on the site within a park, all linked by glazed walkways. The three models survive in the archives at McGill University.

The second scheme best addressed the idea of the museum as a microcosm of a town. The entry pavilion welcomed visitors from the street overlooking the cathedral; a 260-foot-long ramp led across the site and past galleries to the Great Hall, a crystal pavilion, 140 feet high, that echoed the neo-Gothic Library of Parliament, visible across a ravine. To provide shading in the Great Hall, we designed an arrangement of triangular sails extending to form kaleidoscope-like— some say flowerlike—sunshades that diffuse the natural light. From the Great Hall another concourse gave access to more galleries and the curatorial wing.

As the design developed, we faced a decision concerning the 260-foot-long ramp from the entrance to the Great Hall. One idea was to leave the ramp open on one side, with monumental windows overlooking Ottawa and the Parliament complex. Galleries would line it on the other side. The second idea was to place galleries on both sides of the ramp. The two alternatives soon came to be known as the Extrovert and Introvert schemes.

Both approaches had their merits. The Introvert scheme added a certain sense of mystery. As visitors ascended the ramp toward the Great Hall, they would be moving toward an experience that could not yet be anticipated. It also put the focus on the galleries, to the exclusion of other considerations. Many fine art museums are Introvert buildings—such as John Russell Pope's National Gallery of Art, in Washington, D.C. In contrast, the Extrovert scheme put all the museum's public spaces on display to the city, and vice versa.

It was a brash statement about how a public building ought to exist in a pluralistic modern urban context.

The building committee that Boggs had put together consisted mostly of curators, and they gravitated toward the Introvert scheme. They liked the idea that as soon as you entered the building, you would be engulfed by art, and they did not mind the fact that, from the outside, the National Gallery would present a long, blank wall to the downtown area, alleviated only when you arrived at the crystalline Great Hall at the end. I understood the arguments in favor of the Introvert scheme but was truly on the fence when it came to the two approaches. At that point, Jean Boggs said the prime minister had asked that we come to his cabinet with not one but preferably two possible schemes, preferring to have two possibilities to talk about, and she suggested that we bring him the Introvert and the Extrovert designs. I asked if she really had that much faith in a group of politicians, and she smiled.

The ramped approach to Bernini's Scala Regia, Vatican City.

So, we packed up the two models—they were quite elaborate—and went off to Parliament Hill. Six ministers were present for the meeting, along with Trudeau. I made the presentation and the cabinet members started the discussion. They asked if there was a difference in cost between the two schemes. I said the estimates showed that the Extrovert scheme was about 10 percent more expensive—it was less compact and required a vast amount of glazing. The minister of finance chimed in, saying that the Introvert scheme was inherently more Canadian: We are a modest people, he said; we don't show off, don't put ourselves on display. Then the minister of culture weighed in on the side of the Extrovert scheme. This is a new era, he said. Museums are for people, and the transparency and openness represent a revitalization of a museum's traditional role. He invoked Canada's long, depressing winters and envisaged what it would be like to see this crystalline building glowing in the snow. As the discussion wound down, it was clear that the group of six cabinet members was evenly divided.

We had all been standing around the models. Trudeau edged himself behind me and quietly asked which scheme I really wanted. I said I liked both of them—it was much like being in love with two women. And he said, "Yeah, I know." There was a momentary silence, and then Trudeau said, "We are going to do the Extrovert scheme. This is the future. This is what museums will become: open and inviting to the people."

With this decision made, we were on our way. Now we would need to fully develop the design. As an initiation, Boggs, the curators, and I made countless trips to museums around the world to evaluate and see for ourselves innovations that worked and those that did not. Some trips had specific agendas. To give one example: our plan, as noted, called for a 260-foot interior ramp running the full length of the National Gallery. It had to rise a total of about fifteen feet by its end point, so the slope would rise about 5.5 percent—the accepted standard for a comfortable ascent. But Jean was concerned: 5 percent might be fine in theory—but she wondered whether seniors would be put off by the 260-foot length. She wondered whether there was someplace on the planet that already had a ramp of comparable length and slope—something we could walk on ourselves and physically put to the test. We discovered that there was a very old one, at the Vatican: the ramp leading up to Bernini's Scala Regia. Boggs in her later years did not walk at a fast clip. But when we visited the Vatican, she went up and down the ramp for half an hour, and finally announced that she was satisfied.

The ramp at the National Gallery, ascending gently toward the Great Hall.

While we were in Rome (and afterward in Milan), we also visited an array of courtyards, most relevantly the tiered cloister of Santa Maria della Pace, by Bramante, to get a better understanding of light and proportions, because the National Gallery was going to have two courtyards at the heart of the Canadian and European galleries. The use of natural light in the galleries was, from the outset, a major concern. We wanted to bring daylight into all the exhibition spaces, hoping to light the artwork naturally to the extent possible. Given the immense need for wall space to hang paintings, and the fact that sidelight is never ideal in a gallery, exterior windows are not a good option in a museum, and there's also a legitimate worry that the light will prove too strong and therefore damaging. Daylight is easy to provide on the upper floor of the building—skylights have been used for centuries. But with the Canadian collection occupying the first of two floors, Jean was hoping we could somehow find a way to bring the daylight in down below.

We had a eureka moment: we could penetrate the top floor with shafts, six feet wide, and line the shafts with mirrors. The light would bounce down, we hoped, and at the bottom would be almost as bright as it was on the top floor. The client team was impressed but cautious, concerned that it might work in theory but not in reality. I wrote earlier about the importance of physical models—how useful they can be, even if they also have certain drawbacks. Sometimes it is essential to expand the idea of "model" to something larger. To test the idea of using shafts as a way of importing daylight, we built a full-size mock-up of this natural-lighting system—wood, mirrors, glass. For me, mock-ups are an essential tool. They involve creating a piece of a building upfront, in order to refine details and avoid mistakes. I remember once designing an elegant series of windows, discovering only in the mock-up that the screws in the frames were visible from a distance and marred the effect. In the case of the National Gallery, the shafts performed as we hoped, and the idea was endorsed. Someone proposed that the concept should be patented. To that end, we scoured the records to see whether anything like what we had designed had been developed elsewhere. The only possible parallel we found, and it was by no means exact, was a Russian system for shafts to bring daylight unto underground war rooms. During the Second World War, the Russians had constructed subterranean command posts,

Mirrored shafts (top) bring light to the lower galleries; a full-scale mock-up tested the concept.

Fireworks over
the Ottawa River
mark the gallery's
opening, 1988.

Fireworks over the Ottawa River mark the gallery's opening, 1988.

and they wisely took precautions by building in redundancy. The mirror-lined tubes would provide illumination even if electric power were lost.

We expanded the idea of mock-ups again when it came to the Great Hall. To get the siting on the promontory, as well as its height above the ground, just right, we built a wooden platform atop the bluff at the spot where we believed the Great Hall should rise. Boggs and I, along with several others, climbed up to observe the view and determine the precise ideal height from which to see Parliament and the river. To say that we were blown away has a literal edge to it—it was windy up there.

Boggs was willing to put her foot down when she needed to. One issue we disagreed about was the configuration of the floor in the Great Hall. I had proposed a stepped floor, as in an amphitheater, that would lend itself to dinners, concerts, and other events. Jean insisted that a flat floor would provide much greater utility and flexibility. Thirty years later, I can see that she was right.

As we began bringing the design process to completion, one cloud appeared on the horizon: political uncertainty. Trudeau had been prime minister for a total of fifteen years and faced growing opposition. After a reflective "walk in the snow," he would eventually decide to step down. Given the political climate, Trudeau wanted to get the National Gallery to a point of no return, so that a new government would think twice about canceling it. For that reason, we started digging and building the foundation even before the drawings were completely finished. Shifting to construction mode on a project of this scale and complexity requires coordination of personnel and hardware on an almost military scale: the general contractor, the subcontractors, the consultants, the many kinds of equipment, the day-by-day, hour-by-hour sequencing. To watch a project rise brings a sense of achievement but also of heavy responsibility. For me, the transition from *designing* to *making* also produces feelings that I admit to with embarrassment. Thousands of workers are turning into reality what may have started life as doodles on a piece of paper on an airplane.

Ultimately, power did indeed shift to a new government—to Brian Mulroney and the Progressive Conservative Party—but by then the National Gallery was taking form on the promontory. And, in the end, Mulroney embraced the project and was the person who cut the ribbon. The one unhappy development was the departure of Jean Boggs. She had been a close ally of Trudeau's, and as a result of his exit, her position became increasingly fraught. She had always valued her autonomy and now found herself beset by a bureaucratic chain of command she had to accommodate and answer to. It was a sad moment. The gallery staff was beyond loyal—it was affectionate—and Boggs's departure was deeply felt.

The opening of the museum, in 1988, was an extraordinary national event. Trudeau was no longer prime minister, but before the official ceremonies, he and I spent a couple of hours walking through the building by ourselves. I could see in his smile how proud he was. Michal and I hosted a festive dinner in the Great Hall for four hundred people—not just for friends, family, and close associates but also for the many people who had played a role, large or small, during the five-year saga of design and construction. The night of the public opening, one could hardly move in the building. Coupled with the music and lavish fireworks was a palpable feeling of national pride—which is not a typical Canadian form of public display.

With Michal, as the National Gallery of Canada began its public life, 1988.

One of the most moving moments I have experienced as an architect occurred during the week the National Gallery first opened. After ascending the ramp toward the Great Hall, I stood with Michal at its center. A woman and her young son, all of six years old, walked by us. The boy looked up and asked his mother, "Does God live there?"

I do not live and die according to the whims of critics. Many seem fashion-prone, swept up by the mood of the moment, whatever that mood happens to be. Few rise to the level of a Lewis Mumford, Ada Louise Huxtable, or Wolf von Eckardt, with their clear principles and enduring insights. The members of the general public, voting with their feet, turn out to be pretty good critics. Time is the best judge of all. But time does not own a newspaper.

The architecture critic of the *Toronto Globe and Mail*, Canada's most important newspaper, was not susceptible to the National Gallery's charms. Adele Freedman's reaction was critical, and I present a few passages to remind myself of the reality that, in architecture, as in other creative endeavors, the reception of one's work can sometimes come as a slap in the face.

> So what do we have? A low, L-shaped building linked by overhead bridge to a separate curatorial block, the whole a puzzling mosaic of splash and whispers, cellular and modular, shine and drab. Simply laid out along two converging main axes, conceived as a microcosm of the city, the gallery is neither easily legible nor city-like. From the outside, it presents itself as a large, granite-clad box with an overlay of vestigial historical references—a decorated shed that has too many accessories in some places, but is underdressed in others. . . .
>
> The problem is lack of coherence, the absence of an over-all concept in control of the parts. Bold gestures notwithstanding, the National Gallery is a fence-sitter. It's neither urban and tight-to-the-street, nor resplendent on stately grounds. . . .
>
> It's as though Safdie, a vociferous critic of postmodernism, couldn't bring himself to overt symbolism, resorting to coy figuration, instead. That's why the Great Hall appears both bombastic and limp; there's not enough story to it. While Safdie, without question, is a general who can get a project moving and

*win the confidence of his clients, he's been outmaneuvered by a
splendid, textured site.*

The newspaper's editor in chief felt compelled to write a
balancing piece in the same issue. Anthony Lewis of the *New
York Times*, who attended the opening, wrote admiringly of
the building. So did Paul Goldberger. I couldn't help wondering if this could be the same man who had slaughtered
me on the Columbus Circle project. Goldberger understood
what we were trying to achieve. He wrote:

> *The large-scale museum—the museum as civic monument—is
> a central building type of our time, and its problems have
> rarely been solved as thoughtfully as at Ottawa. Mr. Safdie
> has created a castle of glass, concrete and granite that possesses
> as much showmanship as any major museum of the last
> generation, but here the sense of extravaganza never gets in
> the way of the art. There is a remarkable balance between the
> architecture of spectacle that characterizes the public spaces of
> this building and the architecture of repose that characterizes
> the galleries: at Ottawa, it is possible to be exhilarated one
> moment and pensive the next.*

Over the years, the National Gallery has demonstrated
its capacity to be an extension of the public realm of the city.
Its open design and the character of its public spaces have
accommodated scores of state, civic, and social events—all
this as a complement to its primary role as a place for experiencing art. Brian Mulroney began holding state dinners in
the Great Hall, and Prime Minister Justin Trudeau, the son
of Pierre, does so today. The first dinner he hosted was for
the North American Leadership Summit, in 2016. I watched
with great emotion as Trudeau, Barack Obama, and Enrique
Peña Nieto of Mexico walked slowly up the long, red-carpeted
ramp toward the Great Hall.

* * *

In the biblical story of Joseph, years of plenty are followed by years of famine. My experience in Canada has
been exactly the opposite. The National Gallery of Canada
and the Quebec Musée de la Civilisation opened up many

opportunities, and they would continue to arise: the Montreal Museum of Fine Arts; Ottawa City Hall; the Vancouver Public Library; the Ford Centre for Performing Arts, also in Vancouver; and Terminal 1 at Lester B. Pearson International Airport, in Toronto.

At this time, Mexico also became an important part of our lives. In 1989, Michal, our daughters Carmelle and Yasmin, and I had taken a last-minute vacation, joining my brother in Puerto Escondido, Oaxaca. He had been going there for years. At breakfast on a terrace overlooking the beach, we met a couple sitting next to us—Rodolfo and Norma Ogarrio. Rodolfo, it turned out, had been in Ottawa just as the National Gallery was opening. The following day, he invited us to join him to hike three miles down the beach to a promontory that jutted out into the ocean, Punta Zicatela, crowned at its tip by a lighthouse. Iguanas perched on the surrounding rocks, watching us suspiciously. Rodolfo, leading a group of Mexican families, was negotiating to buy the land. They were hoping to build a vacation commune and had recruited a talented Mexican architect, Diego Villaseñor, who was building with traditional methods and was a disciple of the master Mexican architect Luis Barragán. To our surprise, Rodolfo invited us to join the commune; we would be the only non-Mexican family.

Ottawa City Hall, completed in 1994; now the Canadian foreign ministry's headquarters.

And we did. Our house, Balcon, overlooking the bay, would become the place where all of our family would converge—children, grandchildren, partners. The friendship with the Ogarrios has become lifelong. Over the years, Mexico became the center of another important friendship; the architect Richard Rogers and his wife, Ruthie, and their family made an annual excursion to the nearby town of Huatulco, leading to annual get-togethers. The pleasure of Zicatela lies not just in nature's extraordinary beauty but also in the respect for culture and craft. The houses are built from local materials. They are cool without air-conditioning. Zicatela is where I go for contemplation and focused work. I remember laboring on many of the new Canadian projects there for days on end.

The Ottawa City Hall opened its doors in 1994. The existing building, on a beautiful island site in the ceremonial heart of the city along Sussex Drive, was a classic of modernist architecture—a nine-story structure of gray limestone that had been built in 1958 by the man who had been my dean at McGill, John Bland. The challenge was to use the open space behind it and build an addition that had its own powerful integrity—and yet to build it in such a way as to allow Bland's building to live in complementary autonomy. I designed an addition that grew out of the island—one part of

Notebook sketch of the Vancouver Public Library from above, 1992.

it literally rises out of the river, a visual echo of the Château de Chenonceau, in the Loire valley. The river here is the Rideau, which flows as a waterfall into the Ottawa River—a spectacular setting. The new city hall had a glazed arcade running from the entry pavilion to the council chambers. And it paid great respect to Bland's original building by creating on its axis a series of courtyards from which the original could be viewed. Years later, when Ottawa merged with adjacent municipalities and became a new political entity, the city offices were moved elsewhere, and the island precinct was taken over by the ministry of foreign affairs.

The new Vancouver Public Library, begun in 1992, was a prestigious commission. The site had come with significant strings attached. The full city block was owned by the Canadian federal government, and the deal was that, in return for the land, Library Square had to include an office building—meaning a tower—on the site for federal purposes.

Vancouver's mayor, Gordon Campbell, wanted to build public support for the library. He decided that one way to achieve this was to conduct an unusual selection process. It would involve the usual design competition—a short list of four applicants, followed by the submission of design proposals. While these would be evaluated by a jury of prestigious international and local architects, there would be a parallel process. All four designs would be displayed and subjected to a public vote—a kind of referendum. I thought the mayor was taking a huge risk: What if the professional jury differed in its selection preference from the public's? In the end, it

The finished
library, with the
"urban room" in
part of the space
between rectangle
and oval.

The finished
library, with the
"urban room" in
part of the space
between rectangle
and oval.

was not an issue. Our scheme was chosen by the jury and also received 70 percent of the public vote.

The proposal had a number of features that likely tilted opinion our direction. The first is that we managed to integrate the architecture of the office tower and that of the library in such a way that the whole complex had an organic unity. Even now, people often don't realize there is twenty-one-story tower on-site.

The biggest innovation for the Vancouver library was the invention of something we came to call the "urban room." It grew out of the client's program, which stated that the library should have a significant space that was outside the library's actual "control zone"—that is, before one passes through the formal check-in system. Whatever this space was, it had to be public, had to be open eighteen hours a day, and had to offer amenities in the form of shops and restaurants—in other words, a space where people could get together outside the library proper. This space became the urban room, a concept that represented a paradigm change in the way public libraries were used.

That seemingly simple idea led to a prolonged search for the best way to achieve it—a good example of growing a design in response to a program requirement. A traditional library—the New York Public Library's main branch, for example—is usually a rectangular building. In its center is a great reading room, surrounded by stacks. In the case of Vancouver, the client team did, in fact, express a preference for rectangles, citing the efficiency of storing and displaying books. We came up with the idea of inverting the traditional library: placing the stacks and operational components in an efficient rectangle in the center, and then wrapping it with a great oval structure. On one side, the space between the rectangle and the oval provided plenty of room for the reading gallery, facing the city. On the other, it allowed for a seven-story atrium under a skylight. This became the urban room.

The elliptical building was immediately nicknamed the Colosseum. The resemblance in proportion and height was undeniable, and the similarity was enhanced by the low-glow, red-granite aggregate used in the precast concrete building material. Some professional critics sniffed, and some architecture students protested, saying the age of gladiators was behind us. They had a sense of humor—some of the students demonstrated while wearing togas. But the general public seemed to have an immediate affinity for the design. Inaugurated in 1995, it is among Vancouver's most visited sites, by tourists and locals alike.

One feature we had proposed for the library was a roof garden—really, a public reading garden. It was not incorporated into the building because library officials were concerned about security—in particular, an influx of drug users. But years later, after we inaugurated the Salt Lake City Public Library, which featured a successful reading garden on its roof, Vancouver had second thoughts. To mark the twentieth anniversary of the building, in 2015, the city decided to complete the original design by installing the public roof garden at last.

I wish Philip Matthews had been alive to see it. Liverpool-born and -educated, Philip had moved with me from Montreal to Boston and had risen to become the office director. After completing the Salt Lake City Public Library, he joined me on a trip to San Diego, where we were competing for a commission to build San Diego's new library. In the middle of the presentation, Philip suffered a massive heart attack. He was

An *Archie* comic book celebrates the Salt Lake City Public Library, which opened in 2003.

My 1988 cross-sectional sketch for the Ballet Opera House, in Toronto; the project never came to fruition.

only forty-six. I traveled with the coffin back to Boston. It was a shattering moment. Philip had been a rising star, and he would have loved to have seen the final piece of the Vancouver plan put into place.

The library called back the same landscape architect who had worked with us on the original concept—the late Cornelia Oberlander, who was also the landscape architect for the National Gallery of Canada. Her career had many parallels with my own. She, too, was an immigrant to Canada, worked with Louis Kahn, had connections with the Harvard Graduate School of Design, and worked with Arthur Erickson on his many projects. She was ninety-seven years old when we inaugurated the garden atop the library in 2018. (Not long before her death, in 2021, she gave me a call for a discussion about tree selection.)

The inauguration of the garden was a memorable moment. Another, a few years earlier, had involved a different nonagenarian, and the recollection perhaps provides a fitting epilogue to what for years had been a hard-to-unravel relationship with Canada. My life as an architect—like the life of any architect working on an international scale—has had its share of stalwart patrons and determined antagonists. Among the latter was Phyllis Lambert, the daughter of Samuel Bronfman. Phyllis, herself an architect, was a close friend of Philip Johnson and Mies van der Rohe, and a major force. Her initial claim to fame was that she had persuaded her father to hire Mies to design the Seagram Building in New York. She also founded the Canadian Centre for Architecture,

National Gallery of Canada
Ottawa, Ontario, Canada, 1988

Yad Vashem Holocaust History Museum
Jerusalem, Israel, 2005

David's Village, Mamilla district
Jerusalem, Israel, 2009

Vancouver Library Square
Vancouver, British Columbia, Canada, 1995

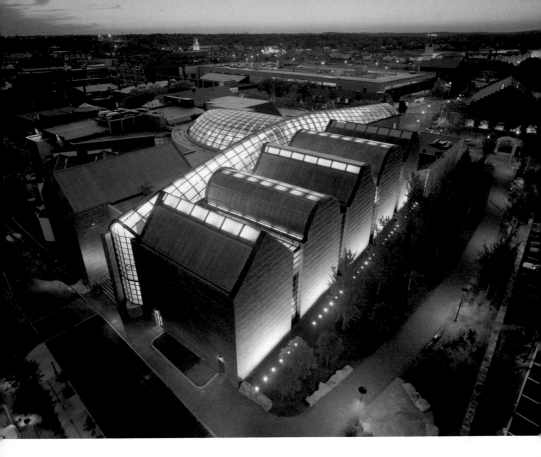

Peabody Essex Museum
Salem, Massachusetts, 2003

Khalsa Heritage Centre
Anandpur Sahib, Punjab, India, 2011

Habitat Qinhuangdao
Qinhuangdao, China, 2017

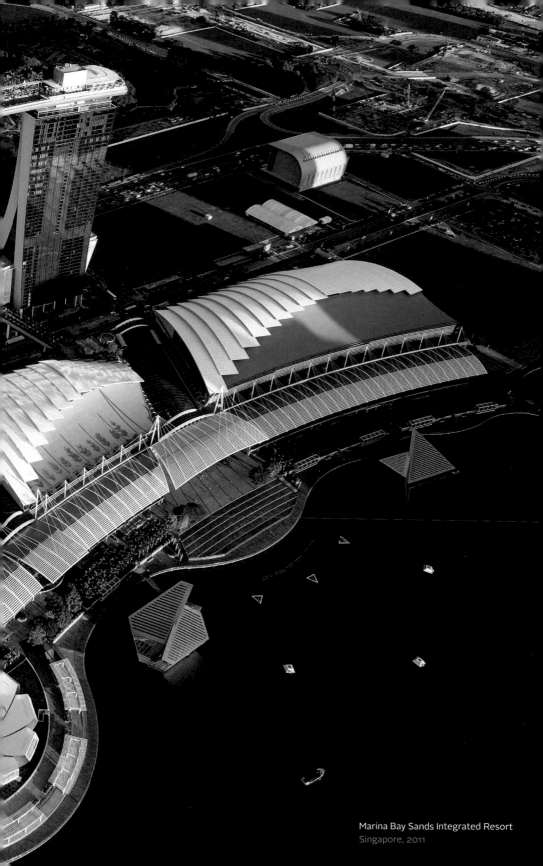

Marina Bay Sands Integrated Resort
Singapore, 2011

Crystal Bridges Museum of American Art
Bentonville, Arkansas, 2011

Skirball Cultural Center
Los Angeles, California, 2013

Kauffman Center for the Performing Arts
Kansas City, Missouri, 2011

Raffles City Chongqing
Chongqing, China, 2020

Jewel Changi Airport
Singapore, 2018

in Montreal, a museum and archive of international caliber that is lodged in a major building designed by Peter Rose.

Phyllis and I had met in 1969, two years after Habitat '67 opened, brought together by a mutual friend. Our first conversation went the wrong way. Untactfully, I had launched into a critique of Mies, drawn largely from my encounters with his residential work in the United States, which I had found so simplistic and dispiriting—and to which Habitat represented a reaction, if not an attempted counterrevolution. Phyllis was ten years older than me, but the real gap was one of sensibility, not chronology. Over the years, Phyllis served on selection or advisory panels in which my work was considered, and she proved to be a consistent, vehement, and energetic opponent (though her views did not prevail).

In 2017, Canada Post launched a series of special stamps. One of them celebrated 150 years of Canada's independence. Another celebrated the fiftieth anniversary of Expo, and an image of Habitat '67 was chosen to represent it. The official unveiling of the Expo stamp took place at Habitat, now leafy and mature, and Habitat residents were on hand for the event.

To my surprise, so was Phyllis Lambert. She was then ninety. I was seventy-nine. Do we mellow with age? I don't think there is a general rule. Some people harden. What I remember from the occasion is that Phyllis unexpectedly came over and gave me a hug. I thanked her for coming and returned the embrace.

The Holocaust
History Museum,
Yad Vashem,
Israel.

CHAPTER SEVEN

Cutting through the Mountain

B uilt structures fall into many categories. They range from the most utilitarian, such as a residence or warehouse or factory, to those that call for an emotional charge—projects that have significant symbolic meaning, the way a parliament building or a place of worship does. Perhaps the most demanding of all are memorials—buildings of mourning, remembering, meditation. Here, the architect is called upon to reach for the sublime.

Do not misunderstand me: a measure of spirituality has a place in any building. Architecture of any kind should uplift the spirit, should make us feel that we are—or want to become—better people, should make us aspire. But in places of memory and places of national and historical significance, something more is needed. To make a musical analogy: it is the difference between marching-band music, rhythmic and energetic, and great piano or cello sonatas or operatic arias or cantatas that reach for the soul.

The most challenging and symbolically demanding project I've ever undertaken is the Holocaust History Museum at Yad Vashem, in Jerusalem—Israel's official memorial to victims of the Holocaust. My introduction to Yad Vashem was a commission I received early in my career, in 1976, to design a museum at the site dedicated to the two million children who perished.

I am a Sephardic Jew who was born in Palestine before Israel existed as a nation. As a boy in Palestine and later in Israel, I of course knew about the Holocaust—it was ever-present in our psyche—but our family had no direct connection with continental Europe. I have nevertheless been immersed in the Holocaust for most of my life because it shaped the lives of the people closest to me.

In my childhood, the constant concern of my German-born nanny, Batsheva, who arrived in Palestine in 1939, for her parents back home gave the Holocaust a sense of immediacy for me. Both of Batsheva's parents died in the camps. Nina, my first wife, was a Holocaust survivor. She was born in the city of Kielce, Poland, and was three when the war broke out. Her father enlisted in the Polish army, and after Poland's defeat fought with Polish forces under Allied command. To escape the ghetto, Nina's mother took advantage of the fact that she and her daughter did not look very Jewish and could both pass as

Gentiles. They hid in the forest for a time. Later, Nina and her older sister lived with a Polish farm family, who put them up in return for a promise of money sometime in the future. The farm family saved their lives, but the conditions were rough. Nina and her sister were confined for two years to a cramped, dark space under a stairway. By the end of the war, they could hardly see and needed surgery to be able to walk again.

My wife Michal, born in Jerusalem—a Sabra—experienced the Holocaust through her mother, Vera Bischitz. Vera was among that massive wave of Hungarian Jews who were shipped to the camps toward the end of the war. She survived only because of the controversial Kasztner deal. Rezso Kasztner, a lawyer and journalist, and a leader of the Budapest Aid and Rescue Committee, had discussed a swap with Adolf Eichmann: ten thousand trucks from the Allies in return for the release of thousands of Jews. While nothing came of that particular proposal, Kasztner did get a trainload of Jews from Bergen-Belsen into Switzerland. The family story has always been that priority went to people whose last names began with the letters A and B. Vera Bischitz and her parents arrived safely in Switzerland in 1944. As a fourteen-year-old girl, Vera kept an extraordinary diary, which I have read. Here she describes her arrival at the Bergen-Belsen camp:

> Everybody had to take off their clothes, and everything, everything, had to be handed in. Not even a ribbon could stay in our hair. We were really in a gas chamber, but thanks to good God, two minutes before the first twenty women went in, an order was received that this transport is not meant to be killed. So, from the gas showerhead, instead of gas, water came. It was my turn. I stripped totally, handed over everything. Then a German soldier (they just wandered around amongst the naked women) took hold of my arm and pushed me into another hall, where young Polish men and women pushed around women and children. . . . I stood in front of a twenty-year-old stone-hearted man. He yanked me closer to himself and pulled my head down to look at my hair. Then, thanks to good God, somebody addressed him, and while replying he kicked me ahead.

As a memorial site, Yad Vashem first opened in 1953. Two decades later, its director, Yitzhak Arad, approached me with the idea of designing a museum devoted to children killed in the Holocaust. In preparation, I spent a great deal of time at Yad Vashem, considering various sites on the campus and also immersing myself in the Yad Vashem archive. One can imagine what the archive has collected pertaining to children. There are letters and drawings and diaries and toys. There are homemade board games. There is clothing. The more one experiences, the more shattering it is.

Vera Bischitz Ronnen, Michal's mother, a survivor of the Bergen-Belsen camp, in the early 1960s.

But the larger context of adding a children's museum at Yad Vashem had to be taken into account: Was it in fact a good idea? There was already a history museum at Yad Vashem, built in the 1950s. The proposed sequence suggested that, after a visit to the history museum, people would then proceed to the children's museum. My concern was that having a second museum that seemed to replicate the first one, the only difference being a focus on children, would be too overwhelming—the visitor, emotionally drained, might not be able to absorb any more. What was needed, I thought, was to memorialize the children—a place of contemplation, not a place of more information.

As I looked around the hilltop site of Yad Vashem, which occupies the ridge of the Mount of Remembrance, I saw what seemed to be a natural rock archway. In fact, it was the opening to a collapsed cave, and I thought we could dig out under the hill, creating a large chamber as a place of reflection.

From the outset, I was thinking of a memorial candle in the space, but rather than the single flame in Yad Vashem's Ohel Yizkor memorial hall (literally, "tent of memory"), I wanted somehow to have an infinity of candles floating in space. As I thought about this idea, I recalled the Labyrinth installation in the pavilion of the National Film Board of Canada at Expo 67. To connect the screening chambers in the pavilion, the designers had created a twisting path of semi-reflective glass that bounced any light source into a thousand versions of itself, seeming to extend outward toward the horizon in all directions. As an experiment, we created a makeshift model with semi-reflective freestanding panels, surrounded by mirrors on the walls, floors, and ceilings. When we placed a candle in the center, the lick of flame exploded into a galaxy of flickering lights. We then built a more careful scale model of the proposed chamber at Yad Vashem, four or five feet in diameter.

To gauge the effect, one could stick one's head inside, covering one's upper body with a dark cloth to keep light from seeping in, like an old-fashioned photographer. I envisioned that, from the cave entrance, people would walk into and across a dark space on a floating walkway, with only the railing to hold on to. The initial sensation would be of nothingness. But soon, as their eyes adjusted, they would glimpse the millions of lights, all from a single candle. And then, at the end, they would exit into sunlight.

The entrance to the Children's Memorial, underground at Yad Vashem.

I was deeply committed to this design. But the building committee at Yad Vashem was disturbed by the proposal. They worried that ordinary people were not accustomed to making sense of abstraction, and they also worried that an arrangement of lights of the kind I proposed would make people think of a discotheque. They decided to shelve the idea. For my part, I had no desire to push back. With respect to the Holocaust, memories and emotions simply run too deep. They must be respected.

A decade went by. The Vietnam Veterans Memorial, designed by Maya Lin, was built, in Washington, D.C., and inaugurated in 1982, and its obvious power may have begun to persuade some of the skeptics about the merits of an abstract memorial. One day, in 1985, Yitzhak Arad, still the director of Yad Vashem, called me again. The Los Angeles banker and philanthropist Abraham Spiegel and his wife, Edita—Holocaust survivors who lost their two-year-old son on the selection platform at Auschwitz—had just been to his office. They had seen my model for the Children's Memorial and were deeply moved, indeed shaken. They offered to pay for the project if Yad Vashem decided to go ahead with it.

So, the Children's Memorial was revived. As we went forward, we had important decisions to make. For instance, we knew what one would see as one moved through the memorial. But what would one *hear*? We needed something that would support the contemplative quality of the space. I recruited Avi Chanani, the director of music programming for Israel state radio—a man who knew everything there

was to find in the station's vast libraries. He started pulling
out records for me to listen to. One of them was a chant by
Paul Horn, recorded inside the Great Pyramid, creating an
echoing sound that seemed to originate deep in the earth.
It seemed perfect, and that became the soundtrack. Com-
plementing the soundtrack is a slow reading of the names
of many of the children killed in the Holocaust. The reading
is done by several voices. Name, age, place of birth. English
and Hebrew. That is all. A nonstop whisper in the darkness.
The Children's Memorial opened in 1987, and Abe and Edita
Spiegel held on to my arms as we descended into the darkness.

There is a coda to the story of my early experience at
Yad Vashem. Only a few years had gone by when, in 1991, the
director reached out to me once more. The museum had just
received a railway car from the German government—one of
the freight cars that carted people to the death camps—and
the question now arose: How to display it? A similar car had
also been given to the United States Holocaust Memorial
Museum in Washington, D.C., and I had seen it there. It
was part of a static exhibit within the galleries, and the presentation did not leave much of an impression. I thought
that a railway car would be best shown outdoors, in the open
landscape, and that it should feel in motion. Walking around
the vast hillside, I thought of placing railway tracks, as if
emerging from the mountain, leading to a bridge. I decided to
cantilever what looked like a vintage European steel railway
bridge, arcing out of the hill over the wadi below, cutting off
the bridge midway, as if it had been destroyed. At the tip of
the cutoff arc, I placed the railway car.

Inside the
Children's
Memorial, the
reflection of a
single candle
multiplies
infinitely.

At Yad Vashem,
a freight car
used in the Nazi
deportations to
the death camps.

Nearby, on the base of the wall of the bridge, words are chiseled in English and Hebrew. The testimony of one of the passengers who survived the journey is haunting.

> *Over 100 people were packed into our cattle car. . . . It is impossible to describe the tragic situation in our airless, closed car. Everyone tried to push his way to a small air opening. I found a crack in one of the floorboards into which I pushed my nose to get a little air. The stench in the cattle car was unbearable. People were defecating in all four corners of the car. . . . After some time, the train suddenly stopped. A guard entered the car. He had come to rob us. He took everything that had not been well hidden: Money, watches, valuables . . . water! We pleaded with the railroad workers, we would pay them well. I paid 500 Zlotys and received a cup of water—about half a liter. As I began to drink, a woman, whose child had fainted, attacked me. She was determined to make me leave her a little water. I did leave a bit of water at the bottom of the cup, and watched the child drink. The situation in the cattle car was deteriorating. The car was sweltering in the sun. The men lay half naked. Some of the women lay in their undergarments. People struggled to get some air. And some no longer moved. . . . The train reached the camp. Many lay inert on the cattle car floor. Some were no longer alive.*

On a handrail there is a poem by the Israeli poet and Holocaust survivor Dan Pagis, titled "Written in Pencil in the Sealed Freightcar," a poem deliberately cut short:

here in this carload
i am eve
with abel my son
if you see my other son
cain son of man
tell him that i

* * *

It has always taken me more effort to realize projects in Israel than elsewhere. The Mamilla district in Jerusalem, for instance, proved to be forty years of agony before being brought to a successful conclusion. The yeshiva project at the Western Wall should have been a warning. There is more controversy in Israel than there is in other places—it is part of the national character. There is also generally less overt expression of appreciation in Israel, or maybe it just comes more grudgingly. Whenever I go to a concert in Israel, even when it is an extraordinary performance, I always find the applause lukewarm. For the equivalent event in Boston, there would be rapturous expressions of appreciation—standing ovations. But as the Hebrew phrase goes, *lo mefargenim*, which loosely translates as "We do not cheer our own."

A number of invited competitions in Israel over the years also ended in disappointment. The first important one, sponsored by the Rothschild family, was for the Supreme Court of Israel. In terms of my own work, I think it is one of the more successful designs I've produced, but it did not make the short list. (The winning Supreme Court design was by my friends Ada Karmi-Melamede and Ram Karmi, who took a very different approach than I had and produced a very fine building.) Years later, I participated in the competition for the National Library of Israel, also sponsored by the Rothschild family, and again did not make the short list. I did, however, win the competition for the National Campus for the Archaeology of Israel, which has recently been completed.

My work at Yad Vashem and, later, at Modi'in, on the coastal plains, signaled an emerging new phase in my life as an architect in Israel. While Yad Vashem is *in* Jerusalem, it is not *about* Jerusalem, and it is not in the heart of the Old City, where so much of my attention had been directed. Modi'in,

too, was not about Jerusalem. It was a city being created anew. Ben Gurion International Airport is another example of this new phase. It is not a gateway to Jerusalem; it is a gateway to the entire nation of Israel.

I had never designed an airport before, but when I learned, in 1994, that a new international terminal at Ben Gurion was being planned, I set about laying the groundwork in order to have a shot at participating. As one who has had to spend much of my time flying, I have been both enthralled and left despondent by airports. They are something of an obsession, and the obsession is not simply about architecture. Transportation buildings have long been a major component of the public realm. In the nineteenth and twentieth centuries, railway stations were treated as flagships of national identity—think of Grand Central, in New York, or Saint Pancras, in London, or the Gare de Lyon, in Paris. Airports today are of even greater significance. One's experience of an airport is about the space, the light, and the clarity, but also about the service, the amenities, the efficiency of passport control and security. I have come to dread any trip through Atlanta: the impossible security lines, the perpetual transfer from one pier to another, the no-man's-land of the public spaces. I single out Atlanta, but I could have mentioned almost any other American airport. The defects are fatal—and that is before even considering what the facilities look like as buildings.

Adding to this catalogue of deficiency, most airports have grown willy-nilly over the decades, the way hospitals do—some new structure is added to an existing one, which itself was once a new structure augmenting an older one, and on and on. The chaos is held together by endlessly zigzagging pedestrian tubes or by light-rail systems that add one more nerve-fraying layer to the act of arrival and departure. This process of accretion made a quantum leap in the 1960s and '70s. The result was inevitable. Airports everywhere—Frankfurt, JFK, Heathrow—succeeded in banishing comfort, reason, efficiency, and delight.

To have a chance to design the new terminal at Ben Gurion meant researching a subject I understood only as a frequent passenger. It meant getting to know the two or three key people in government who had thought deeply about the airport project, were responsible for pursuing it, and were willing to speak frankly about their ideas and about what kind of airport—and architect—they wanted. Laying the groundwork also meant being more intentional in terms of my own appreciation of the airports I visited—trying to understand what worked and what didn't.

When it comes to designing airports, certain qualities need to be front of mind. An airport is an interface, and the whole point is to get passengers, as fast as possible, from the point of arrival to the door of the airplane. Everything in between is an unfortunate necessity. Minimizing the in-between for the ordinary traveler is a challenge, because, over time, speed bumps have been added—for security, for passport control, for shopping—and also because the volume of traffic is so high. Another bump comes in the form of hubs; whatever the rationale for them, at hub airports, some 70 percent of the people in the terminal at any given moment are waiting for connecting flights.

Ideally, passengers should never feel lost inside the airport. Clearly, and intuitively, they should be able to know exactly where they are and where they need to go, no matter where they're from or what language they speak. There are many ways to provide direction, to nudge people along. The architecture should be self-orienting. Don't create mazes, don't box people in. Daylight—through skylights and windows—can persuasively

The courtyard of the National Campus for the Archaeology of Israel, completed in 2019.

suggest the proper way. Sometimes destinations can be made obviously visible in the distance.

When buildings were smaller, offering a clear sense of orientation was more easily accomplished. A museum such as the Frick, in Manhattan, is an elegant mansion with a courtyard surrounded by rooms; the Guggenheim, also in Manhattan, has a spiral ramp that similarly gives access to all the galleries. But the Metropolitan Museum of Art, with as many public rooms as days of the year, and the British Museum, whose sprawl mirrors its collection, on which the sun never sets, are a different story entirely. To make navigation through such behemoths easy is a challenging task. It is all the more so when "wandering" is not an option—at airports, say, which are even bigger than museums and where visitors have to get to a specific space at a specific time. As buildings get larger, adding more of the same does not work.

Large buildings must be thought of as mini-cities, with the equivalent of main arteries, landmark buildings, side streets, parks, and identifiable smaller destinations. It's a lesson we derive from many Greek and Roman cities: the north–south-running *cardo maximus*, the main spine of the city, intersecting with the east–west-running *decumanus maximus*, the other major thoroughfare. These monumental, arcaded streets provided guidance no matter where one was from.

In a related point, the design of airports needs to reflect the fact that these facilities are today, effectively, urban spaces. They are not simply places of transition; they have some of the characteristics of neighborhoods and commercial districts. Airports that serve as hubs are holding tens of thousands of passengers captive for hours. One solution is to provide opportunities for shopping, and there is an enormous incentive to route passengers through retail precincts and relieve them of their money. The conflict between the convenience of the passenger and the financial interest of the airport is clear. Balancing them is the challenge.

Further, an airport must manage large numbers of people who are in a state of stress—low, moderate, high, or off the charts. People are stressed about flying, about making their flights, about security, about customs, about the meeting they're going to, about whether they left the stove on at home. So, an airport should be soothing. It should be full of light and air and landscaping. It should be soft.

An airport is also a place where thousands of people work. To neglect them and the circumstances of their working life is to ignore one of the most important clients—the "silent clients," the ones who rarely get a seat at the decision-maker's table, even though they will spend the largest portion of their lives, aside from time spent at home, living with those decisions. Concern about their well-being should not be limited to places visible to passengers, such as checkout counters, but must include the areas known as the "back of the house," where workers change their clothes and eat their meals and take their breaks and perform the countless unseen tasks that keep an airport alive—areas that are usually drab and windowless.

Finally, there is the matter of security. The security requirements at an airport are complex and demanding. At Ben Gurion, the philosophy behind them dictates that it's better not to be aware of the security. Passengers rarely see a group of uniformed soldiers with machine guns, as they do in some airports—a visible show of force. But there are many security people one never notices. And at Ben Gurion, a visitor does not see that the sheet glass is laminated to prevent lethal shattering and does not notice the thin floor-to-ceiling cables meant to keep panes from dislodging in a blast—preventive measures tested by U.S. military laboratories.

Designing a terminal at Ben Gurion proved to be an education. What airport authorities usually do when embarking on a new terminal is hire specialist master planners to come up with a basic diagram—a kind of conceptual framework. For instance, some airports have a central hub from which individual piers radiate outward. Some are shaped like a U, others like an E or an H. Some have several satellites linked by trains. Some have separate pavilions for check-in and baggage claim, and some incorporate those two functions into a single building along with everything else. Years ago, Eero Saarinen devised an innovation for Dulles International Airport, just outside Washington, D.C.: a traveling lounge that rolls from the gate directly to the airplane, eliminating the long walks—an innovation full of potential but never embraced by either airlines or passengers.

The overseers of Ben Gurion had hired master planners who proposed a particular scheme for the new terminal: it would have a "landside" building, where departing passengers check in and arriving passengers pick up their baggage, and it would also have an "airside" building, starting at passport control, linked by a connector to a rotunda where passengers could shop and await their flights. Five piers, radiating outward, would lead passengers to their gates.

When it came to the process of selecting the architects, Israel's Ministry of Finance laid down a set of rules for the competition. The challenge was immense: the old airport had elements dating back to the British Mandate and was perpetually overcrowded; passengers still walked onto the tarmac for many flights. The new airport—airside plus landside—would be quadruple in size and cost around $1 billion. Because no airport of this scale had ever been built in Israel, the ministry wanted Israeli architects entering the competition to team up with a foreign firm with airport experience. The ministry also decided that putting the entire airport project into the hands of one team was too risky. Instead, one team would do the landside and one team would do the airside. I decided to go for it and collaborate with TRA (now Black & Veatch), which had built many airports around the world. A second team, led by my friends Ada Karmi-Melamede and Ram Karmi, joined forces with SOM, another global firm with airport experience. They won the landside, and we won the airside. The teams worked closely together from the start.

We designed the roof of the rotunda—the main area where passengers wait—at Ben Gurion as an inverted dome that channels rainwater to a central oculus at the dome's bottom, where it drops into the waiting area as a circular

A sketch of Ben Gurion International Airport, 1995. The inverted dome collects rainwater, and a waterfall tumbles from the oculus.

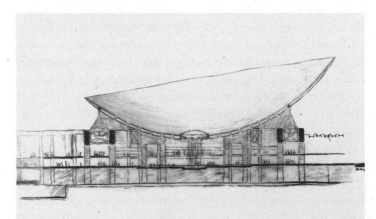

waterfall. For reasons that are perhaps bound up with our emergence as a species, the sound of moving water—the splash of a fountain, the babble of a stream, the rhythmic crash and retreat of waves—performs a kind of magic for the soul: sometimes soothing, sometimes energizing. I have made use of water in almost all of my projects.

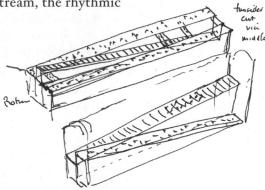

The master planners had proposed an arrangement typical of many international airports, where departing passengers move through primary concourse levels from the time they step onto the sidewalk until they board an airplane. Arriving passengers, in contrast, are sent down into the lower level, underneath the terminal above, deprived of views and daylight, to passport control and then to the less-than-inspiring baggage hall. Only then can they make their way outdoors for a first glimpse of the country that is supposedly welcoming them. Fly into Heathrow's Terminal 4 or New York's JFK, and this is the arrangement all passengers find.

Arriving and departing passengers, visible to and criss-crossing one another. A notebook sketch, 1995.

This standard procedure is unfortunate anywhere and especially so in Israel. About 99 percent of all visitors to Israel arrive through Ben Gurion—there are not many other options—and for many first-time visitors and immigrants, it is an emotional event. So, I proposed inverting the traditional flow. Arriving passengers, rather than descending, would go up one level, cross the airport in a light-filled mezzanine, and only then head down for passport control and baggage claim; from the higher reaches of this walkway, they would have a view of the surrounding landscape and also of the departing passengers below. There would also be a point along the connector where arriving and departing passengers would crisscross—on two ramps, facing each other but slanted in opposite directions, separated and yet visible to one another. To give the setting gravitas as well as local flavor, the surrounding walls were made of Jerusalem stone. On the ramps, the sheer prominence of the movement of people in two directions underscored the idea of Ben Gurion as the nation's gateway.

* * *

One prolonged—three-decades, all told—project in Israel that, in the end, I can look back on and say was completely worth the effort is the city of Modi'in. In 1988, I was called in by the minister of housing, David Levy, and the director of the ministry, Amos Unger, to create a master plan for a new city. My work on the Jewish Quarter and the ongoing plans for Mamilla were well known, and in this instance, I was not competing for the commission but participating in an effort that had started out almost as speculation. The new city Levi and Unger envisaged was to be located halfway between Tel Aviv and Jerusalem, on the Israeli side of the Green Line. To make an important point explicit: In the West Bank, there have been many Israeli developments, known as settlements, across the Green Line. On principle, I have not participated in any such projects.

The initial impetus for Modi'in was the ever-rising cost of housing in both Tel Aviv and Jerusalem—to a point out of reach for young couples in particular. The matter became more urgent in the late 1980s and after, when the emigration of Jews from the Soviet Union (and then Russia) became a flood tide; Russian Jews came to Israel by the hundreds of thousands. Building the new city now became a project of high national priority. Modi'in was to be a major city of 250,000 people on an eight-thousand-acre site (Tel Aviv's population, by comparison, was 435,000), and bringing it to fruition would involve not only devising a master plan but also coordinating and supervising the work of dozens of architects and developers over a period of twenty-five years. In the design of a new city, there are multiple locational issues to consider—climate, transportation, traffic, density, topography, drainage patterns, subsoil conditions, the state of the aquifer—as well as issues of lifestyle and culture. Provisions must be made for schools, clubs, libraries, and other institutions. One needs an awareness of the underlying economics, land values, and infrastructure costs.

One also has to cast an eye at other significant planned communities. A number of them have been built in Israel, such as Beersheba, Ashdod, Arad, and Carmiel, and famous cities have been built internationally, such as Brasília, in Brazil, and Chandigarh, in Punjab, India. There are many new postwar towns in the United Kingdom and Scandinavia. The new towns in Israel were conceptually based on the British model—a grid of autonomous neighborhoods, each with its own center, and arranged in a supergrid of arterial roads. Much has been written about the shortcomings of this model; cities that have grown organically develop more like a tree, with a trunk and radiating branches. On their own, they do not come to resemble sheets of graph paper inhabited only by the tidy citizens of *The Truman Show*. And as we all know from personal experience, neighborhoods have facilities of various kinds—as we desire—but they are anything but autonomous.

Theories and diagrams, however brilliant, mean nothing unless the site is understood. A plan for Tel Aviv, on flat dunes, doesn't have to deal with a varied topography, but the prevailing winds that come from the sea have to be taken into account (and worked with). Modi'in was a virgin site with undulating hills and valleys, and not many trees, but it was not a tabula rasa. It had been the home of the Maccabees in the second and first centuries BCE and later the site of Byzantine and Ottoman towns. The ground holds much from antiquity, and archaeologists have been active over the years. I remember touring the site for the first time in a Jeep, taking it down into the streambed on the valley bottom. That in itself told me something important about making the valleys the spines—the lifelines of the town. The army had built a dirt road down there for maneuvers. At one point I stopped the Jeep to avoid hitting a turtle on the road. I got out, picked it up, and moved it to the side. I remember thinking: This is but a temporary reprieve; something big is coming. Scouting the terrain also brought home a larger truth. This was not the kind of site that would accept the imposition of some theoretical idea of what a new city should be like. One could not simply impose an abstract diagram upon it. One had to grow the plan with the shape of the land in mind. So, to start with, we re-created the terrain in the form of a large and detailed model, consisting of fifty pieces, each of them a little more than a yard square. We needed a space the size of a two-car garage just to house it.

I always thought that the best parts of Tel Aviv, a city designed by Patrick Geddes in the 1920s, were the tree-lined boulevards that crossed the city, leading to the sea, with double rows of trees running along the wide median. People strolled up and down, stopping at small kiosks and cafés.

The boulevards are similar in certain ways to Las Ramblas, the famed promenade in Barcelona.

Modi'in's valleys are oriented more or less east–west, flowing toward the sea and capturing the prevailing winds. We decided to put the neighborhood roads as pairs in a one-way pattern on each side of the valley bottoms, and between the roads we placed parks, schools, synagogues, shops, and other community facilities. Taking a leaf, so to speak, from Beverly Hills, we planted each valley with a distinctive kind of tree, creating the valley of the pines, the valley of the palms, the valley of the jacarandas. The place where three valleys converge became the town center. The valley that ran from the town center and widened toward the sea, Wadi Anaba, was set aside as a nature preserve. (Indeed, 50 percent of the valley space in Modi'in is parkland.) With the basic framework of the plan in place, we contained the edges of the other valleys with housing, which starts on the flat edges and then extends up the hills, in terraced formation.

We wrote urban-design guidelines and other regulations into the plan. Tall buildings were confined to the hillcrests, so that they did not overshadow the housing lower down. We did not allow developers to "cut and fill"—leveling the terrain to make building cheaper—as developers like to do. They had to follow the topography. All residential units had to have access to a garden, a terrace, or a balcony—the lessons of Habitat. We did not design the individual buildings, but buildings had to conform to the guidelines. When an architect had a plan for a particular sector, she or he had to make a model and then insert it into our much larger model, like a puzzle piece, so we could see how it fit with the whole. The various tracts designed by some twenty architects, all slotted into place in a larger model of a neighborhood, led to inevitable conversation. We didn't have to force improvements—they became obvious to everyone.

Critics in Israel like to pick on Modi'in. It's supposed to be a city, they say, but it's not really urban. It's just a place where people sleep. Back in 2001, just when Modi'in was officially declared a city, Esther Zandberg wrote in *Haaretz*: "Even though the city was promoted as a 'city of the future,' the planning principles of Modi'in make it a quintessential suburban development, a bedroom community whose liveli-hood depends on external jobs. Today, the city only has 500 jobs." Give it time, I have said. Tel Aviv was not really "urban" in 1935. Twenty years ago, Modi'in began as a dormitory town. Now, there are almost as many jobs in Modi'in as there are working residents—it has become a magnet. The population has surpassed one hundred thousand, and it continues to grow. Modi'in has a mayor and a city council, and it runs its own municipal affairs. And twenty years from now, it will have acquired a true urban character. It will have evolved into a city organically.

Our experience with Modi'in would carry over to a new project undertaken in the mid-1990s—the development, at the behest of Jerusalem mayor Ehud Olmert, later Israel's prime minister, of a master plan for the region that lies just west of the city limits, the only direction in which Jerusalem can reasonably be expected to grow. The official assumptions for the master plan had stipulated certain densities. I agreed

Modi'in as it has actually developed.

to take it on only on the condition that I and my own team be allowed to determine the appropriate densities—and, further, be allowed to consult with environmental groups throughout the process. These conditions were agreed to. The planning process began by mapping those areas where, in our judgment, development was out of the question. This involved some subjective considerations—for example, a decision that valleys should remain open green space, forming a continuous necklace of parks to serve the city's future population. To this we added the archaeological areas, often next to areas with natural springs. After defining the green framework, we were left with the residual lands available for development.

The master plan ultimately ran into lethal opposition from many quarters. To our utter surprise, the green groups, with whom we had been working all along, came out against it, claiming that implementation would destroy Jerusalem's green lungs, when in fact it would preserve them. Meanwhile, politicians and settler groups on the right opposed making plans for westward development, because they wanted Jerusalem to develop to the *east*, encroaching deliberately on potential territory of a Palestinian state. Opponents began calling the master plan the "Safdie plan," as if it were something I was personally trying to impose on Jerusalem rather than developing as part of an official planning effort launched by the city itself. The West Jerusalem master plan went nowhere. For good measure, I was burned in effigy by demonstrators.

* * *

In 1995, almost a decade after completing the Children's Memorial, we participated in the competition for rebuilding the history museum at Yad Vashem. The existing history museum was out of date and overcrowded—the annual number of visitors had grown since the 1950s from three hundred thousand to more than three million. In terms of substance, the narrative presentation of the Holocaust required more detail than the old space could hope to provide. Vast amounts of material released from Soviet archives had become available, which there was no room to display. Meanwhile, the opening of the U.S. Holocaust Memorial

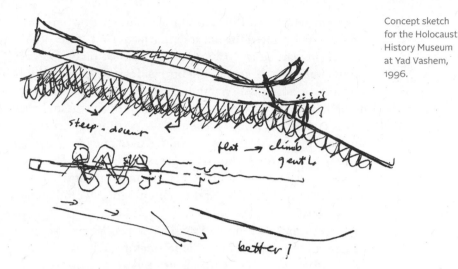

Concept sketch for the Holocaust History Museum at Yad Vashem, 1996.

Museum, in Washington, D.C., had in many ways made the older museum at Yad Vashem obsolete.

I had by then walked every inch of the site, among the beautiful pine trees and rocky escarpments of the Jerusalem hills. The new program, established by Yad Vashem's building committee, called for 180,000 square feet of narrative exhibits as well as a gallery for Holocaust art and a synagogue—a very large building. Locating it was the challenge. My first thought was that the Holocaust didn't need an architectural monument—or, perhaps more accurately, that no such monument could rise to the challenge of the Holocaust. The Holocaust is an enormity. It cannot be represented by a conventional work of architecture. In a sense, I was wondering what kind of space might possibly give primacy to the Holocaust's own terrible reality. I felt from the outset that a new museum should not be a building per se—should not represent business as usual. It should not be a massive structure squatting atop a pastoral hilltop. With the experience of the Children's Memorial in mind, I began to think about underground quarries, carved out of the bedrock—nothing associated with architecture as we conventionally think of it. At some point, a strategy emerged: we should build the museum underground. As I envisaged it, a long, linear structure, triangular in cross section, would penetrate the mountain from the south, cutting through like a spike toward the north and bursting out over a valley and toward the light and the trees. There would be no presence at all for a building other than where the structure penetrated the hill and then jutted out at the other end. Inside, along the way, would be exhibition chambers carved out of rock, straddling both sides of the prism-like structure—individual chambers, each unfolding a chapter in the narrative. Deep cuts into the floor of the prism held artifacts; the cuts would prevent visitors from shortcutting through. Visitors would have to move right and left, zigging and zagging through the chambers, seeing light at the end as they proceeded but able to reach it only by following the narrative in its entirety. Natural illumination would come from above, from skylights.

The proposal seemed to send tremors of concern down the spines of those responsible for making a decision—the selection jury and the board. They opted for a second round, asking the competitors to elaborate, and then a third. Eventually they decided to entrust us with the commission.

Once chosen, we faced a number of reality tests. The initial thought had been to carve the space for the new museum out of the bedrock and leave the rock as the natural wall of the galleries. But the rock turned out to be too soft, and there was too much water leaching through. Eventually, we opted for lining the walls and floors with concrete. The museum would also have to be able to handle large crowds, including many elderly people and people with disabilities. In this respect, had the museum been built at the top of a hill, it would have posed difficulties—requiring escalators or elevators. Our concept overcame the problem, since by going through the mountain, the museum was kept more or less on a single level from entry to exit. Getting from one end to the other was a matter of ramps and grades rather than stairs.

The completed museum, a shaft penetrating the hill.

Yad Vashem hired an exhibit designer, the late Dorit Harel, with whom I had worked before, on the Skirball Museum at the Hebrew Union College in Jerusalem. She brought all her great skills to bear on the task, despite the fact that, unbeknownst to us, she was coping physically with a severe degenerative disease. We collaborated so as to entwine the historical narrative with the evolving plan of the galleries. We built a twenty-foot-long model of the entire museum. It was large enough to allow Dorit to display scaled-down exhibits, photos, cases, and artifacts. We could modify the size of the chambers and adjust the lighting. In the museum, as one experiences it, black-and-white photographs merge into the gray concrete surfaces. Physical objects find appropriate niches for display. As visitors go right and left and across, they wind through the whole tragic history—the rise of anti-Semitism, the ghettos, the camps. We also manipulated the floor levels. At the outset, the path descends gently, on a 5 percent slope. Visitors go deeper into the mountain, feeling as if they're slowly sinking. Toward the end of the narrative, the slope reverses and starts

The view overlooking the Jerusalem forest at the far end of the Holocaust History Museum.

to rise. Climbing, visitors can see at the prism's end a distant sky—but only sky, because of the upward slope. Only when they reach the tip of the prism does the view open suddenly before them, exploding outward: hills, trees, neighborhoods. We prevailed. We are here. Life continues.

This concept of the emergence into the light overlooking the city was controversial. It took months of discussion to get there. Some of the ultimate decision-makers worried that the architecture was dominating the narrative—that it was too optimistic and too celebratory. But I believed, passionately, that the architecture made an appropriate statement: that it focused, sharpened, and amplified the narrative, and emphatically did not compete with it or seek to supplant it. Perhaps I was also trying to avoid the architecturally ambiguous conclusion to the Holocaust museum in Washington—an inward-looking octagonal hall that leaves the visitor suspended, almost deserted. In the end, the curators, historians, and survivors on the committee were persuaded by our design.

The new museum opened in March 2005, in the presence of forty heads of state and other world leaders. Early

With Barack
Obama, then a
senator, at Yad
Vashem, 2007.

in the day I walked through the museum with Elie Wiesel, experiencing it through his eyes. That night, Wiesel spoke powerfully about the inadequacy of understanding and the paramount necessity of memory: "We go through the museum and we don't understand. All we know is what happened. And now the question is: What does one do with memories? Any psychiatrist will tell you, if you suppress memories, they come back with a fury. You must face them. Even if you cannot articulate them, we must face them."

Years later, I received a call from our longtime friend Samantha Power, a journalist and foreign correspondent with a deep commitment to the cause of human rights; and, later, the U.S. ambassador to the United Nations. (She currently heads the United States Agency for International Development.) Samantha had left a position at Harvard to go work for a relatively unknown young senator from Illinois named Barack Obama. "My man is about to enter the presidential race," she said. "He is traveling to Iraq and Israel. Would you travel to Jerusalem and take him through Yad Vashem?" It was a low-key visit, which took place in 2007. He had only two people with him. An hour was scheduled. The visit extended beyond two hours. What impressed me was Obama's total concentration. One expects politicians to occasionally take a phone call, send a text message, look at a watch. Obama did not. This was no obligatory visit; it seemed to me a transformative experience. The questions and comments indicated knowledge and compassion.

After Obama left, I called Michal: "I just met the next president of the United States," I said.

* * *

Official guests in Israel are often taken to Yad Vashem as a matter of course. One of those visitors, in 1997, was Parkash Singh Badal, at the time the chief minister of Punjab. Israel and Punjab had established a number of cooperative programs, mainly in agriculture, and these were the formal purpose of Badal's trip to Israel. After seeing the Children's Memorial, which affected him greatly, Badal asked if it would be possible to meet the architect.

As luck would have it, I was in Israel. I received a call from the foreign ministry, asking if I would be available. A grand old man, very tall, with a long white beard, Badal wore his hair beautifully wrapped in a white turban. He had big farmer hands but spoke with a soft, gentle voice—he was a priestly, almost biblical, figure. Badal was accompanied by two armed guards, also in turbans, when I met him in his hotel suite. He said to me, "We Sikhs have suffered a great deal in our history. We've been persecuted, like your people. We have fought for our survival. And now we are going to build a national museum, to tell our story. I would like you to come and design it for us." The new Sikh museum was to be called the Virasat-e-Khalsa—the Khalsa Heritage Centre.

It is always gratifying when someone is genuinely moved by our architecture. But I have also developed defenses to take some of what is said on such occasions with a grain of salt. Sweet expressions may be ringing in my ears, but a commitment hasn't been made. And yet two weeks later, I was on my way to India. From the moment of my arrival at Chandigarh, Punjab's capital, I was treated like a head of state, with bodyguards, limos, and all.

We first traveled to Amritsar to visit the Golden Temple—the Mecca, or Jerusalem, of the Sikhs. It is an extraordinary manifestation, with its magnificent golden dome surrounded by a grand rectangular pool, which in turn is enveloped by arcade after arcade. Then, from Chandigarh, we drove off to see the proposed site of the new museum, two hours away on rutted, potholed roads. There were Jeeps ahead of us and Jeeps behind, all carrying Sikh soldiers with machine guns. The sudden immersion into the world of the Sikh people brought images of something familiar. Every face—the features, the expressions—reminded me of my own family members.

The Khalsa Heritage Centre was to be built in the holy town of Anandpur Sahib. We visited the fortress Qila Anandgarh Sahib and also the Gurudwara Takht Sri Kesgarh Sahib, the main temple, where in 1699 the guru Gobind Singh formally founded the Khalsa Panth. The proposed site for the museum lay six miles outside of town—as isolated as one can imagine. A practical question immediately arose: How would anyone ever get to the museum? This was rural India.

By now I felt close enough to the chief minister to speak frankly. I questioned the site selection, and we returned to

Parkash Singh Badal, chief minister of Punjab, India, 1998.

Anandpur Sahib to see whether there might be a site in walking distance from the temple and fortress. I had noticed two hilltops not far from the center of town. There was nothing built on either of them, and I pointed to them and suggested we could build on both hills and then connect across the valley. My hosts promised to see what could be done. A few weeks later, I learned that they had purchased the two hills and the valley between them. There had been no opposition from a Municipal Art Society. The speed was surprising, and one can only imagine the backroom dynamics.

And in this case, money seemed to be no object. In most projects, the tension between ambition and budget is keenly felt. In the course of almost every project comes a moment of "value engineering": the moment when the client, working with the architect, reviews every element of the program and design to find places to cut corners and scale back. The phrase is, to my mind, a cynical one, implying that one can add value by cutting costs. Most value engineering results in cutting quality—and clients generally come to regret their decisions. The Khalsa Heritage Centre was different—indeed, it was unlike any other project we've participated in: the project had no real budget. The aim was to produce the best possible building, and in our many meetings the stern gaze of the accountant was noticeably absent. The assumption was that fund-raising from the Sikh diaspora would somehow cover the cost.

I began to learn more about Sikhism. There were already rumblings about an American Canadian Israeli coming in out of the blue to impose some foreign vision. Sikh scholars sat down with me on each of my visits, offering personal seminars. A Delhi-based associate architect—the late Ashok Dhawan—was recommended as someone who could be my collaborator. Ashok and I became friends. Together we thought through our needs for a good general contractor and recommended Larsen & Toubro, a Scandinavian company that had been active in India for many years and possessed a thoroughly Indian professional staff and labor force. Larsen & Toubro had built the lotus-like Baha'i temple, in Delhi.

Meanwhile, I was thinking about the scheme and studying the site. As I began sketching, I found myself reflecting on the fortress architecture that is so much a part of the Sikh tradition—great structures that crown hilltops, the product of human hands and yet organically integrated into the landscape. One sees them everywhere in Punjab and Rajasthan, and they served as a powerful inspiration. I conceived of the museum as clusters of buildings on each of the hilltops, connected by a pedestrian bridge that spanned the valley. The buildings and bridge would be constructed of concrete and clad with the golden-colored sandstone that gives character to so many buildings in the region.

Recalling the grand pool at the Golden Temple and the presence of water at almost every Sikh temple, I proposed damming the valley and creating a series of cascading ponds. The pedestrian bridge arcing across would be reflected in the water. Back at our home office, in the United States, I listened to Punjabi music as I worked on a proposed design.

Weeks later, when I presented the design in Chandigarh, the reaction of those who had some sort of say in the matter—Badal and other Punjab government officials, as well as Sikh religious leaders and a number of scholars and architects—was mixed. I quickly inferred why. The model I had brought, made of Plasticine, was not as realistic in terms of materials and landscaping as a more developed model would have been, and didn't serve any of us well. I found myself continually having to offer reminders: "This is sandstone" or "This is water." For the next trip we made a smaller model, beautifully crafted in wood. The roofs were made of reflective metal, representing the stainless steel we intended to employ. We used mirrors to realistically render the water garden. Trees and vegetation were glued into place.

The heritage center today, evoking a Sikh hilltop fortress.

The Khalsa center grew organically out of the sandstone cliffs. In one building, clustered cylindrical shafts with their tops carved away like inverted domes—the inverse, rendered in silver, of the white domes of Sikh temples—formed a floral silhouette. It is known today as the Flower Building. On a clear day the silhouette of the museum complex stands out against the Himalayas, to the north.

The design was fully embraced by the clients. We were still drawing and developing the design when I returned for the groundbreaking. The bridge was already partly built. To my amazement, hundreds of thousands of people were in attendance, many on horseback, all in festive Sikh regalia. It was clear what great significance this project held. There were horse races and mock battles. For the first time, the groundbreaking at one of my projects was attended by beautifully adorned elephants.

That memory would have to sustain me across some difficult years. The project's first problem was something that I had understood was not a concern: money. The notion that rich Sikhs around the world—and there are many of them, like Didar Singh Bains, the so-called Peach King of California—would be eager to contribute to this noble effort proved optimistic. I traveled with Badal on fund-raising trips to London, Vancouver, and various places in California, but money in the needed amounts was not forthcoming. One concern on the part of some potential donors was that the Khalsa Heritage Centre was a government project; many people of means were reluctant to entrust funds to the government.

Electoral politics also intruded. The museum was half-way built when elections resulted in the Punjab government changing hands. Badal, of the Bharatiya Janata Party (BJP), was out, and the Indian National Congress candidate, Amarinder Singh, an army captain and the maharaja of Patiala, was in. In Canada, I had seen projects hang in the balance because of electoral politics. I didn't suppose that politics was more placid in India. Luckily, my colleague Ashok was politically well connected. At one point he got wind of unsettling rumors—that our project, strongly identified with the Badal regime, would be turned into a hospital. Eventually, with Ashok's help and that of some of the civil servants involved, we secured an opportunity to tour the site with the maharaja. In the end, he not only committed himself to the project but, as Badal had never done, put a line item in the government budget to fund and complete it.

The maharaja was a man of distinction. He lived in the family palace, adorned with the faded trappings of princely entrenchment—Victorian portraits, ornate sofas, and diverse works of art, both Indian and foreign. The maharaja was a historian. Among the books he had written was a history of the Sikh regiments that had fought in the Great War.

Four years later, as the Khalsa Heritage Centre was nearing completion, there was another election, and Badal was back at the helm. He, therefore, got to cut the ribbon. Michal and I met him in Chandigarh on the day of the opening, and we flew by helicopter to Anandpur Sahib. No two-hour drive this time. I had never before seen the museum complex from the air. It seemed as if it had always been there—looking like

With Michal and associate architect Ashok Dhawan (at Michal's elbow) during the opening of Virasat-e-Khalsa, 2011.

an archaeological treasure that was also completely contemporary. It would eventually claim the world record for the largest number of museum visitors in a single day.

On the ground, elephants were again on hand to greet us. Sikhs had done most of the construction, and the pride was palpable. When the throngs were at last allowed inside the museum, they took off their shoes as they went in, as they do at a temple. In their eyes, this was a sacred place.

* * *

When I think about my experiences in Israel as an adult and about the history of the country during the past half century, one moment stands out as a terrible fulcrum: November 4, 1995. That is the day Israeli prime minister Yitzhak Rabin was assassinated. Michal and I were devastated by the news and traveled to Israel for the funeral.

Rabin and his wife, Leah, had become friends over the years. I had met Leah when she came to learn more about our work in the Mamilla district. The visit was organized by the site's developer, Alfred Akirov. He noted that Leah had heard some of the criticisms of our work and wanted to see for herself. The tour went well. As we walked around, Leah mentioned that she and Yitzhak would be visiting Canada, and one of the things she hoped to see was the National Gallery. I arranged to take them through the building myself, traveling to Ottawa for the occasion.

Yitzhak and Leah Rabin, 1977.

Whenever we were in Israel, we made a point of getting together with the Rabins. Often we were guests at the prime minister's residence. They were very different, the two of them. Leah had dark hair and penetrating green eyes, was very outgoing and sophisticated, and enjoyed the finer things in life. Yitzhak was very Israeli—down to earth, a heavy smoker, laconic. He had a good sense of humor. Yitzhak could be described by the biblical expression *melach haaretz*, "salt of the earth."

We hosted a memorable dinner in our house in Jerusalem in October 1994, the evening of the day Israel signed a peace treaty with Jordan, in Aqaba. We had planned the dinner two

months earlier. Yo-Yo Ma, accompanied by Marty Peretz and
Michael Kinsley of *The New Republic*, were coming for a visit,
centered on Yo-Yo's performance with the philharmonic.
Knowing that Leah was an admirer
of Yo-Yo's, we had scheduled a dinner.
Several Israeli friends would join us.
Three days beforehand, we learned
that the peace treaty was going to be
signed that same afternoon. Michal
and I welcomed this good news, and
sent a note to Leah, assuming that the
dinner would have to be postponed.
Her surprising response came back:

"We will be done by the late afternoon. Clinton has decided
that he will proceed to visit Assad after the signing. We will
be free, but will join you half an hour late."

In King Hussein's
helicopter with
Yo-Yo Ma,
Jordan, 1994.

    It is a testimony to the spirit of the time that although
Rabin had to walk five hundred yards from the parking area
of the Jewish Quarter to our house, through narrow alleys, he
was accompanied by only two security agents. In more recent
times, if Benjamin Netanyahu had made the same journey,
there would be an army's worth of security, and the quarter
would be buttoned up tight. The evening was celebratory
and yet surreal. Halfway through the meal a call came in,
and Rabin's security man gave him the phone. It was Egypt's
president, Hosni Mubarak. Rabin walked out to our terrace,
which overlooks the Old City, and spoke with Mubarak for
about five minutes. After he returned, he talked about his
own personal transformation—a military man who had come
to believe that military action would never resolve the Israeli-
Palestinian conflict. He also talked about his conversation
that afternoon with Jordan's King Hussein, joking how they
had had to resort to secrecy to share a smoke and a moment
of intimacy and friendship. Later, when it was time to part,
we walked downstairs with the Rabins and found the narrow
street thronged with people eager to see and greet their
prime minister on a triumphant day.

    We had planned to travel the next morning to Jordan with
Yo-Yo Ma to visit Petra, but with the peace treaty signed, the
trip took on a completely different character. As it turned
out, we would be the first Israelis to cross the Allenby Bridge,
spanning the Jordan River, as tourists. We were then picked

up by official emissaries. After visiting Petra, we unexpectedly ended up at the king's palace in Aqaba for dinner. Sitting on an open deck by the sea with King Hussein and Queen Noor, Yo-Yo Ma performed the prelude to Bach's Cello Suite No. 1 in G Major followed by Ernest Bloch's *Prayer*.

We were in Cambridge on that November evening in 1995 when we received word of the assassination. Michal and I had gone to the Fogg Museum for an event to mark its temporary closing. The architect Renzo Piano was about to rebuild Harvard's museum complex. Amid the happy buzz of the reception came a phone call with the terrible news. We traveled to New York that same night and flew from there to Israel.

A few days after the funeral, Leah called to ask if I would be willing to design Yitzhak's tombstone. She would also ask me, much later, to design the Yitzhak Rabin Center, which is the equivalent of a presidential library—a home for Yitzhak's archives and a place for conferences and special exhibitions. The tombstone was urgent: traditionally, a tombstone is put in place on the thirtieth day after a person's death. There had been a standard design in the plots at the official cemetery, on Mount Herzl, set aside for the heads of state and other top leaders—"the greats of the nation," as they are called. The standard design is just a slab of black lava stone from the Golan, but Leah believed that, since this was an extraordinarily traumatic event, a special design was in order. I would, of course, need to present the design to the Israeli cabinet for approval.

The tombstone designed within days of Yitzhak's assassination, 1995.

Switching from the scale of a building to that of a tomb-stone was a seamless process. Working with our model shop in Boston, we produced a dozen schemes using slabs of white balsa wood and black ebony, representing limestone and volcanic lava stone, both plentiful in Israel. Because this would ultimately be a double tomb—for Yitzhak and, one day, Leah—I expanded on the idea that the lava stone would represent Rabin and the limestone would represent Leah. I thought the tomb should not be a flat slab, but spatial. I began with two cubes, carving a conical section out of each, and then placed them together so that, in combination, they created a small niche or apse. In the center, between them, I placed a memorial candle—in reality, it would be an eternal flame—calling to mind the thousands of candles that had been lit, spontaneously, all over Israel the night after the assassination. I packed up that idea and a half dozen other versions and brought them to Israel. I placed all of them in front of Leah and her daughter, Dalia, and said nothing about my own preference. After about thirty seconds, Leah reached for the design with the candle and said, "This one."

The design still needed to be reviewed by the cabinet. A meeting was scheduled, and I took the model with me. The prime minister, Shimon Peres, was present, along with six cabinet ministers. I explained the idea, and it was quickly approved. I had expected controversy, or at least some debate—this was Israel—but the overwhelming emotion of this moment in the nation's life seemed to have brought out a different attitude, if only for a few hours. An artisan stone-mason whose skills I had admired at Yad Vashem quickly found the appropriate blocks of lava (from the Golan) and limestone (from Galilee) and by hand gave shape to the design.

At the unveiling, emotions were still raw. Mourners filed by Rabin's tomb, placing a stone atop it, in the Jewish tradition. The flame glowed in its conical surround. Sadly, Leah Rabin joined her husband not many years later. It was left to their daughter, Dalia, to complete the Rabin Center. I am grateful for a family friendship that now extends to a second generation.

The Crystal Bridges Museum of American Art, in Bentonville, Arkansas.

CHAPTER EIGHT

The Power of Place

Alice Walton is a remarkable woman, as I would come to know first-hand. Of the four children of Sam Walton, the founder of Walmart, Alice, his only daughter, is perhaps least associated with the running of the family business, though she certainly keeps a hand in. Depending on the fluctuations of the stock market at any moment, she is either the richest or the second-richest woman in the world. She grew up in Bentonville, Arkansas, a quintessential and unassuming small town before it became the headquarters of the world's largest retailer. It is said that Sam Walton considered Alice the "maverick" among his children, like the maverick he himself had been. Alice was involved for a period with investment banking. She is a breeder of champion horses. Her great passion is art. Knowing what great wealth has done to the second and third generations in some families, I have found it impressive—in fact, reaffirming—that she has devoted her life to giving back, to enriching the community she came from.

The centerpiece of her gift to Arkansas and the nation is the Crystal Bridges Museum of American Art, which opened its doors in 2011. The collection was assembled by Alice herself, and includes works by John Singer Sargent, Andy Warhol, James McNeill Whistler, Georgia O'Keeffe, Edward Hopper, Thomas Eakins, Childe Hassam, the various Wyeths, and many others. Alice owns Norman Rockwell's iconic *Rosie the Riveter*. In 2005, she bought Asher Brown Durand's *Kindred Spirits* from the New York Public Library. The museum is located on the family estate, walking distance from downtown Bentonville. Sam Walton's tiny original five-and-dime store—Walton's 5-10—is still present on the square. From its headquarters, nearby, Walmart manages more than ten thousand stores worldwide. The town is served by a regional airport whose purpose, to a significant degree, is to bring people either to Walmart or to Crystal Bridges. So far, the museum has welcomed five million visitors. Alice has been known to give impromptu tours, not revealing her identity. The museum building itself represents a distillation of some of my thinking about the relationship between architecture and nature, and, in turn, between art and nature.

* * *

Back when Crystal Bridges was just a dream, I had heard that Alice Walton was thinking about establishing a museum of American art. She had hired a director, Robert Workman, and we learned that she was quietly visiting museums around the country. Architects talk among themselves, and word gets around. Specifically, we heard that Walton had visited two of our projects—the Skirball Cultural Center, in Los Angeles, and the Peabody Essex Museum, in Salem, Massachusetts—and had toured both of them incognito. One day, in November 2004, Workman gave me a call: Alice would like to meet. Could I travel to Bentonville and spend the day? I got on a plane, after making a quick trip to see a Walmart store—I had never been to one before—and landed at Northwest Arkansas National Airport, which Alice had cajoled the state government into building. She can be persuasive. But she is also one of the most astute and collegial clients I have worked with.

Alice picked me up in a Lexus sport-utility vehicle that she drove herself. She was wearing jeans and said, "I'm going to take you to my cabin to have dinner, and we'll chat, and tomorrow we're going to look at the sites." Her place in the Ozarks, in a forest overlooking a beautiful lake, really was a cabin—rustic, built of logs. There was another gentleman with us, John Wilmerding, a Princeton art historian who advised Alice on art purchases. Walton cooked three steaks. We ate, had some wine, went to sleep. The next day we jumped into the SUV, picked up the museum director, and started making the rounds.

First, we visited the family estate, where the house Alice grew up in is located. Bentonville itself, I saw, consisted of little more than a village square, though it is today far more developed and affluent than any other small town in northwestern Arkansas. The family estate extends across a wooded hill, with a stream and ravine running through it. Crystal Springs, located upstream, feeds the flow year-round. At one point, we walked down into the ravine. There was a big pipe, slippery with moss, laid across the ravine to enable people to walk from one side to the other, though without a railing. The spring was high, running over the pipe, and when I stepped onto it, I lost my footing and plunged into the water. We continued on, my wet pants flapping. It was a cold January day, and I could feel my pants gradually begin to stiffen.

We drove around and visited other sites. We talked all
day: museums, daylight, build here, build there. The dynamics of designing for an individual who has total authority
over every decision are very different from designing for an
institutional committee. It is all one-on-one, two interacting
wills and sensibilities. It is a dance: action, reaction, leading,
being led. Late in the afternoon, as Alice took me back to
the airport, I asked how she was proposing to conduct the
selection process. Would there be a competition? Did she
have a short list of architects? Would there be more inter-
views? She paused, smiled, and said that her search process
was over. She had already done her homework. She had seen
museums by other architects. She had visited my own projects. "I have completed my search today," she said. As she
explained much later, she recognized certain elements in
my buildings—in particular, the respect for natural settings
and the passion I have for creating spaces that people seek
out and want to gather in—that she desired. She wanted the
museum to be more than a museum; it should also be a center
of community life.

The first crucial decision was where to build. I encouraged her to stay within the family estate—the first site we had
seen, where I had fallen into the stream. I initially proposed
a scheme siting the museum on the hilltop overlooking the
ravine, but I quickly realized that to build there would mean
taking down hundreds of mature, beautiful trees. Building
within the ravine instead would follow the tradition of the
old mill towns, built near the streambed, close to mills pow-
ered by water. We could build dams across the ravine, creating ponds as beavers do. Gallery spaces could also straddle
the ravine. We could keep the natural flow running. This
picked up on the design of Walton's family home, built by the
remarkable Arkansas architect E. Fay Jones, where he bridged
the same ravine with the family living room, and also built
a dam that created a pond where one could swim. But I was
also thinking of Frank Lloyd Wright's Fallingwater, where a
stream flows under the house before turning into a waterfall.

From my first sketch—two dams, two ponds, two buildings
that were also bridges across the ravine—ensued a process
that involved many months of investigation and exploration.
Building in a drainage path is complex. We had to deal with
the U.S. Army Corps of Engineers and the Federal Emergency

Management Agency, which are deeply involved in flood control. Initially, the engineers were nervous about the ponds, worried about the accumulation of silt and debris. Their recommendation was to have the natural flow of the watercourse pass underground, below the museum, through large pipes, and then, above, at ground level, create two waterproof ponds of "managed water." These would essentially be glorified versions of swimming pools. We opposed the scheme. Our intention was to preserve the watercourse as it was, its passage among the buildings recalling the flow of water through a town. I personally believed that making the ponds integral to nature's path would prove both effective and economical. And so it has turned out. The ponds are a natural part of the watercourse. There is no dredging, though we do aerate the water to control algae.

We were up to our elbows with the planning process when Michal and I left on a long-planned trip with friends to Bhutan, in the Himalayas. We hiked among the mountains and ridges and, like the locals, crossed the ravines over narrow, handmade suspension bridges. The bridges served people but also cattle. There in Bhutan it dawned on me: Why not stretch cables across the ravines in Bentonville and let the cables support the buildings above the ponds? Upon my return, we worked out the technicalities with Buro Happold, a global engineering firm (London's 2012 Olympics stadium, Beijing's international airport, and our own Jewel in Singapore) and

An early notebook sketch for the Crystal Bridges Museum of American Art, 2005.

one of the best in the business. We had worked with the firm's
engineers before, and they were excited by the challenge.

In the end, we brought the idea to fruition. Crystal
Bridges opened in 2011. It was a radi-
cal assemblage. The bridge buildings
were supported by cables across the
flowing water. For the roof structures
spanning and enclosing all the museum spaces, I believed that laminated
wood beams would provide visual
warmth. Moreover, they were very
sustainable, as they were fabricated
from local Arkansas yellow pine. The
curved and vaulted roofs would be

With Alice Walton,
the museum's
creator and a
visionary client,
2011.

clad with copper. It would oxidize, first to dark brown and
then to a turquoise-green patina. Working out the connections between beams, cables, skylights, and glazing made
Crystal Bridges one of the most elegant buildings we have
ever designed. The ambience as a whole is calm. The water
teems with fish and turtles and everything else in a natural
Arkansas pond. It goes brown after a storm and then settles
back within a few days to crystal blue. It sometimes freezes
in winter. It is alive.

A visitor to Crystal Bridges likely has no idea of the many
thousands of decisions and revisions, large and small, that
underlie the finished museum complex—nor should they.
Sifting recently through some old office correspondence
between the field office and the home office, I came across
a few emails that give a sense of the kinds of issues that
arise. For instance:

> *Tuesday, September 11, 2007*
> *The Building C balcony length has increased by approximately 9".*
> *Buro Happold must evaluate whether the current beam sizes on*
> *the structural drawings can accommodate the additional length.*

And this:

> *Thursday, October 26, 2006*
> *In another location we are interested in using a bottom-up*
> *shade on sloped glazing. If we have an outward-sloping plane of*
> *glass (obtuse angle on the interior, 104 degrees), can the Nysan*

*bottom-up system work with cable guides similar to the skylights
or does it require tracks?*

And this:

Wednesday, September 12, 2007
*I wanted to let you know that yesterday afternoon we
encountered a situation in the rock face of the excavation behind
A2 and the museum store. We were in the process of beginning to
install the rock/soil nailing in the noted areas. When we cleaned
the loose material off, we encountered some deep fissures back
behind the rock that we felt might be caves.*

As we collaborated on Crystal Bridges, Alice and I took
several trips. We visited the artist Jamie Wyeth at his home
in Maine—a house on Monhegan Island once owned by the
artist Rockwell Kent. I also went with Alice to the Louisiana Museum of Modern Art, outside Copenhagen, with its
significant collection of modern art. What took us there
were the conversations we had been having, as the Crystal
Bridges galleries evolved, about experiencing art and nature
together. At the Louisiana Museum, a series of pavilions
extends along the seashore on beautiful grounds dotted with
sculpture. One can walk outdoors, as well as through glazed
walkways from one pavilion to another. Alice was inspired,
hoping to achieve a similar conversation between art and
nature in Arkansas.

When we returned, we reviewed the plans once more,
looking for potential places where we could reciprocate nature's
embrace. At moments the embrace is almost literal—one may
emerge from a gallery of New England landscapes and then
cross a glazed link that opens out to a view of blossoming dog-
woods. Contained spaces give way to the uncontainable. An
entrance lounge to a gallery at the south end of the museum
lies not far from two magnificent trees, nicknamed Thelma and
Louise. We altered the design in order to save them. Nearby,
a deep ravine made straight for the museum. We created a
large glass wall overlooking the ravine, which dips under the
lounge and brings water into the pond. It is a joy to sit there,
watching people experiencing a joy of their own.

There were, of course, difficult moments along the way.
At the outset, we had to deal with contractors who were

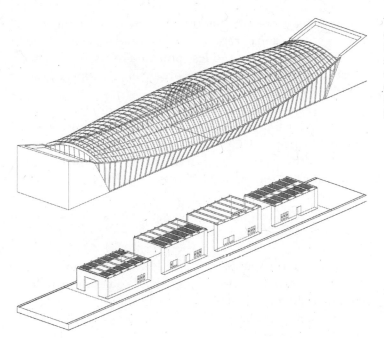

The exterior of one of the bridge buildings suspended over Crystal Springs (top) and the interior galleries it encloses.

convinced that we had designed a building that could not be built with local know-how. For them, much about the design was unprecedented and therefore difficult or impossible to realize. Walmart construction managers were invited to become involved; their experience, however, substantial as it might have been, was with a different kind of building entirely. At one point, the original director, Bob Workman, left, and a new director, Don Bacigalupi, came in. The atmosphere changed as soon as he arrived. He had little trust in architects and seemed to view everything through a lens of suspicion. Though we were more than halfway through construction, he campaigned against having skylights in the galleries. As we had shown at the National Gallery of Canada, nothing compares with seeing works of art in natural light, and it can be done safely, without harm to the art. But Alice was persuaded to cancel the skylights, something she has since had second thoughts about.

All this aside, Alice had a sophisticated appreciation for the work, and she understood the nature of the site. She was, and is, an extraordinary client. Alice knew what she wanted and at the same time was open to talking about what she may not have known she wanted. And she was willing both to trust and to take risks. When we built a mock-up, we tested a sleek-looking concrete, mixing together gray and white cement, and compared it with standard gray cement, which is more rustic and less polished in its appearance. Most clients would prefer the more-manicured option; Alice voted for the earthier solution.

As construction was coming to completion, a contest was launched to name the museum. Many suggestions ensued, but Alice's proposal, Crystal Bridges, appropriately turned out to be the winner—"Crystal" for Crystal Springs, the water source, and "Bridges" for the galleries suspended over the water.

* * *

The key to the success of Crystal Bridges was unleashing the power of place: unlocking what I sometimes call "the secrets of the site." Unleashing the power of place had also been the key to Yad Vashem. And to Modi'in. And the Khalsa Heritage Centre.

Ancient builders have long understood the power of place. They have built monuments and towns and homes with this in mind. From distant eons of prehistory, they have been aware of the way one can leverage location to accomplish what a built structure cannot achieve on its own. Sometimes the location is spectacular. Think of the ancient city of Petra, carved into red sandstone canyons; think of the surrounding environment, the wind-whipped wadis and djebels of the Jordanian desert. Or think of utilitarian structures like the Roman aqueducts, coursing on tiers across impossible valleys, their extraordinary beauty as architecture and engineering informed by the dictates of the land.

Or consider Machu Picchu, in Peru. The Incas could have stayed in the plains by the coast—and, indeed, they did build temples there. But in Machu Picchu, they found a place where nature has produced one of its wonders. And then they stood on nature's shoulders—its nearly inaccessible shoulders—so that architecture could reach higher. They achieved something that they could not have achieved on their own.

Unlocking the secrets of the site has applications that go beyond the world's monumental places. It applies to ordinary buildings in ordinary locations. In places where hills were surrounded by fertile ground in the valleys, farmers built settlements on the slopes to conserve the usable land. When defense was a priority, sites were chosen that could be walled or protected by a river or ravines or by the sheer fact of height. Building on sloping land, the builder took advantage of the terrain by terracing structures, maximizing gardens and views for all. That is the "secret" behind the Italian hill towns and the beauty of the Amalfi Coast and the alluring ordinariness of Hadar HaCarmel, my old neighborhood in Haifa.

One of the books I treasure, and that has influenced my thinking, is Bernard Rudofsky's *Architecture without Architects*, published in the early 1960s as the catalogue for an exhibition at the Museum of Modern Art, in New York. The book celebrates the so-called vernacular builders of almost every civilization. Rudofsky wags his finger: "Skipping the first fifty centuries, chroniclers present us with a full-dress pageant of 'formal' architecture—as arbitrary a way of introducing the art of building as, say, dating the birth of music with the advent of the symphony orchestra."

Leveraging location: Machu Picchu, Peru.

Whichever of these two paths is involved—the monumental or the vernacular—the proper stance is to work with nature, not to seek to defy or transform it, to approach nature with humility, as a student who has much to learn. This stance defines not only what architecture should do; it defines the approach we should take to managing our planet. Deciphering the secrets of the site involves considerations of topography, hydrology, vegetation, altitude, sunlight, and climate. In urban areas, it involves an awareness of surrounding buildings. It can even involve individual trees, as was the case with a federal courthouse we designed in Springfield, Massachusetts.

A generation ago, the U.S. General Services Administration, which is responsible for constructing all federal buildings and had a well-deserved reputation for mediocrity, decided to start pursuing excellence in architecture. The new direction, embraced in the 1990s, was very much driven by Senator Daniel Patrick Moynihan, of New York, and by the GSA's chief architect, Edward A. Feiner. An indefatigable builder with the buzz cut of a 1950s football coach, Ed Feiner created a program called Design Excellence and invited a wide range of American architects to compete for commissions. Richard Meier, Thomas Phifer, Henry Cobb, and Thom Mayne all designed courthouses or other buildings. Feiner didn't have a stylistic or ideological ax to grind; he just wanted work that was good. Under his eye, the federal government put up more buildings than at any time since the Great Depression. Our firm ended up winning three competitions—two of them for regional federal courthouses and one for the headquarters of the Bureau of Alcohol, Tobacco, Firearms and Explosives, not far from Union Station, in Washington, D.C.

The site of the Springfield courthouse, on State Street, lay at the heart of a historic city. During the American Revolution, George Washington had made an encampment

U.S. Federal Courthouse, Springfield, Massachusetts, which embraces historic trees.

right up the hill. In the middle of the site stood two giant trees, more than 200 years old—a beech and a walnut. Others participating in the design competition for the new courthouse proposed taking the trees down. When I went to the site and saw the trees, I was struck by the fact that they were as old as the American union. They likely had another hundred years of life in them. To me, these trees were sacred, and I wasn't going to cut them down. So, in our proposal, which the GSA embraced, we left them intact and designed a curved courthouse spiraling around them. We consulted a tree expert to define the area that had to be left free in order to protect the root systems; it was a radius of forty or fifty feet. With that established, we designed the building as a crescent that held the trees and a garden in the embrace of its curve. The interior of the curve—where the building's circulation spine would be—was almost entirely glass, giving people inside on every floor a view of what had been saved. This gesture of homage to the trees was celebrated by designing a grand staircase that ascended toward the courtrooms by following the crescent. The trees were the centerpiece. Everything rotated around them.

* * *

By the time we started work on Crystal Bridges, we had been intensively involved in designing and building cultural institutions in the United States for about a decade—starting with the Skirball Cultural Center, in Los Angeles, in 1986, which Alice Walton had visited when looking for an architect to work with.

Toward the end of the 1980s, as we were completing the series of successful Canadian projects—from the National Gallery to the Vancouver Public Library—I realized that it was time to be moving on from intense involvement with Harvard and teaching. I had been in Boston for nearly ten years. Things had gone well. At Harvard, I had been appointed the Ian Woodner Professor at the Graduate School of Design. Because I was both running the department and teaching, the time commitment was great. Increasingly, I was aware of the conflict between my growing practice and the responsibility I owed to the students. Also, I continued to feel somewhat isolated in the school. I seemed out of sync with many members of the faculty. For a variety of reasons, it seemed like a good time to turn the page. In the Harvard community, to resign an endowed chair is unusual. I remember Harvard's president, Derek Bok, who had by then become a friend, calling me to ask if my resignation was a protest. I told him, quite sincerely, that it wasn't. It was just time to turn the page.

But, truthfully, I was disappointed. While I had the appreciation of students and the satisfaction of seeing them thrive in the world, my satisfaction did not extend to the school as a whole. My interests—indeed, convictions—did not coincide with those of most faculty members. The calendar would eventually turn on postmodernism, in part because much of what it produced was superficial and cartoonishly grotesque. But the ethic of permissiveness remained. It was followed by a series of other "isms." Today, when a young architect comes into the profession, he or she is exposed to a perplexity of formal choices and motivations. If I try to put myself in the place of an aspiring architect today, I come face-to-face with a confusing swirl of influences and models: exuberant and sculptural: Frank Gehry and Zaha Hadid; high-tech and elegantly crafted: Renzo Piano and Norman Foster; minimalist: David Chipperfield; environmentally responsive: William McDonough and Ken Yeang.

Some of these people are friends. Much of the work, and the ingenuity, I admire. But consider the sheer diversity as it must appear to a young set of eyes. Trying to find a satisfying answer to the questions of what one is supposed to be doing, and how to develop a value system, is difficult.

On site with Uri Herscher (left) and the late Jack Skirball (center), 1985, forces behind the Skirball Cultural Center, in Los Angeles.

Meanwhile, new doors were opening. The first institutional project I designed in the United States—the Skirball Cultural Center, in Los Angeles—turned into a decades-long relationship. Its origins lay in an earlier project. In the late 1970s, I had designed the Hebrew Union College campus in Jerusalem—a school with roots in Reform Judaism—and in the process had come to identify strongly with the institution. The chair of the board, Richard Scheuer, had been one of those dream clients: a New York philanthropist whose interests ranged from yachting to biblical archaeology. One of the people I met through this association was the dean of Hebrew Union's Los Angeles campus, Uri Herscher. At the campus he oversaw a small museum of Judaica, the Skirball Museum, which had a long history and had been given new life (and a new name) by the movie mogul Jack Skirball, the man who produced Alfred Hitchcock's *Shadow of a Doubt*, among other films. Jack had grander ambitions, and saw the museum as the nucleus of what could become a cultural center with wide reach. Herscher, an ordained Reform rabbi as well as a scholar, took on the task of turning Skirball's ambition into reality. We had much in common, Uri and I. We were virtually the same age, had both been born in the land that would become Israel, and had both migrated with our families to North America. We shared basic values.

The path to collaboration on Skirball had been difficult. The board had at first insisted on having a California architect, and I had enthusiastically suggested my close friend David Rinehart. David teamed up with another Louis Kahn alumnus, Jack MacAllister, who had been the project director for Kahn's landmark Salk Institute. I stayed on the sidelines as an adviser. On paper, it looked good. But David's gentle soul and Jack Skirball's Hollywood brashness did not mesh. As tensions worsened, Uri turned to me. For obvious reasons, I was reluctant to get involved, but in the end, I felt compelled to grasp the nettle. I leveled with David about the situation. He was magnanimous and generous, and urged me to take over. But he was deeply disappointed and pained. So was I.

A site had been secured, a fifteen-acre tract abutting Sepulveda Boulevard and Interstate 405, a little to the north of where the new Getty Center would be built. But whereas the Getty crowns a hilltop, the Skirball site lay in hardscrabble lowland. In fact, it was a garbage dump.

I studied the site, which curves around the base of a hill where the Santa Monica Mountains meet I-405. Small ravines create inlets running into the hills. It was crucial to understand all the risks involved, the usual California hazards: fire, earthquakes, mudslides. California's morass of regulations could also make one dizzy. With all this in mind, we began to think about how a campus, including a museum, a conference center, an outdoor amphitheater, auditoriums, classrooms, assembly and event halls, courtyards, and gardens—could not only respond to the challenge of the setting but be animated and empowered by it.

Given the complexity of the site, my mind summoned precedents, such as the Hadrian's Villa complex, in the hills outside of Rome, with its series of pavilions carefully fitted into the contours of the landscape. I had visited Hadrian's Villa during that long European sojourn after my family left Israel and had never forgotten the experience. I also thought about the ancient sanctuary of Delphi, on Mount Parnassus, in Greece, which exemplifies the Greek approach to planning, a subtle dialogue between topography and architecture. Hadrian's Villa had been an exception—the Romans typically preferred everything to be axial, symmetrical. If a mountain

Notebook sketch, 2013, for Skirball, at the foot of the Santa Monica Mountains.

was in the way, it was moved or leveled. The Greeks were different. If the topography was an impediment, then the building was adjusted accordingly. If some other "fault" defined the site, the architecture conformed. Greek building sites often have no rigid plan, even though the form of each building remains rigidly classical. That interplay is what I was after here, though the forms of the buildings would owe less to the Greeks and more to the modernism-inflected California vernacular of Richard Neutra and Rudolph Schindler. I also concluded that, given the lifestyle and climate of Los Angeles, every programmed indoor space at Skirball should be complemented by a corresponding space outdoors.

Defined by the shape of the land: the remains of Hadrian's Villa, in Tivoli, Italy.

Many of these outdoor rooms, which take up roughly 50 percent of Skirball's footprint, are gardens, including sculpture gardens, but the outdoor space also consists of courtyards, discrete in character, partly enclosed by the buildings and partly open to the hills so that nature flows right into them. Some have water. Some are paved but shaded by jacaranda trees. There are outdoor spaces for concerts, for weddings, for more intimate gatherings. Some features serve a double purpose that is not obvious: the outdoor amphitheater, for instance, also serves as a retaining wall to guard against mudslides.

The simple palette of the complex is related to that of the Hebrew Union College in Jerusalem. Galleries, classrooms, and assembly spaces are constructed of concrete, each structure surrounded by an arcade containing walkways and galleries, supporting trellises on the upper level. Each cluster of columns in the arcade is crowned by a large planter with wisteria and climbing roses, extending horizontally to cover the trellis. With Skirball I sought to achieve a seamless integration of landscape and architecture, and I came closer to success than I ever had before. The roofs were vaulted, with clerestory skylights integrated into them, lighting the spaces below. Because visitors almost always viewed the building with the dark Santa Monica Mountains rising behind, we decided to roof the vaults with stainless steel—reflecting the sky and marking a contrast with the dark backdrop. An inlay of pink granite panels was set into recesses in the concrete walls to provide a warm, soft glow.

I also collaborated with Michal's mother, Vera Ronnen, an enamel artist working at architectural scale, who had created murals at Jerusalem's Hebrew Union College. Uri Herscher invited her to do the same at Skirball. This became an expanded family affair. Our daughter Carmelle, who studied art at Cooper Union and Bard, was coming into her own as an accomplished artist, and she joined forces with Vera. The two of them spent time together on the Isle of Wight, where large-enough kilns were available at a factory. Vera's deep knowledge of enamel thus skipped a generation to Carmelle. I often find myself wondering—and marveling—about the transmission of talents and values within families and down the generations, and by what pathways it occurs. I grew up in a community-minded setting in Israel and now have the pleasure of seeing our daughter Yasmin become a passionate—one might say radical—community organizer and social worker. She is the director of programs at a nonprofit that, among other activities, creates visual political-education tools that break down complicated policy. My son Oren, as noted, is a playwright who trained as an architect. My daughter Taal is an architect. She and her husband, Ricardo Rabines, have worked together on the West Coast since 1993, running their own very successful firm, Safdie Rabines. Taal, Ricardo, and I had the good fortune to collaborate on one major project: the Eleanor Roosevelt College, at the University of California, San Diego.

Skirball today, between hills and highway.

Many visiting the Skirball Cultural Center have referred to it as an oasis. Skirball was built in four stages over a period of twenty-five years. Jewish history and philanthropy gave it context and life, and continue to do so, but it has become a place for all of Los Angeles. On weekdays, school buses bring children from every part of the city, paid for by the cultural center. The friendship that developed between Uri Herscher and me, and the mutually respectful relationship with the board, have made working on Skirball a joy. We never signed a formal contract; we did not need one. I have watched with admiration as Uri built Skirball into an institution that is central to the region's civic life. Skirball also marked a new chapter in terms of the evolution of my architectural language, drawing on Jerusalem and the California vernacular. My sketchbook, a reliable Rorschach test, is relatively thin on Skirball—there were no angst-ridden dark nights of the soul, registered by page upon page of rejected drawings. Instead, there was a self-assuredness and a feeling of familiarity from the outset.

* * *

Skirball was the first of the big institutional commissions we received in the United States. We started getting more, mostly as an outcome of invited competitions, another of which was for the Peabody Essex Museum, in Salem, Massachusetts.

Salem had been one of the most important seaports in Colonial America, and in 1799, a group of civic-minded ships' captains, who had plied the seven seas, decided to create a museum—the first one in the country—to display the many "natural and artificial curiosities" they had acquired. The original East India Marine Society Hall, an elegant classical building, still stands as part of the current museum complex. Several additions were built in the ensuing centuries as the marine society evolved into the Peabody Museum, with important holdings that ranged from Asian, African, and American art to Colonial-era furnishings to photographs to ship figureheads. In 1992, after the merger of the Peabody Museum and the Essex Institute, the board of the new Peabody Essex Museum decided to build a major new addition and undertake a total reorganization of the museum's physical plant. An international competition was organized, with four firms invited to participate, ours among them.

It was paramount that the new design result in a clear and cohesive overall organization, somehow bringing order to the helter-skelter collection of existing historic buildings. The scale of the surrounding city presented a contextual challenge: how to add a structure of one hundred thousand square feet (as the museum stipulated) in a way that did not disrupt the delicate scale of the eighteenth- and nineteenth-century houses that formed much of the neighborhood—an urban fabric that was as remarkable and compelling as a tract of beautiful landscape. We decided to break up the mass of the new building into a sequence of five parallel pavilions with pitched roofs and brick facing that resonated with the surrounding domestic architecture. Skylights along the peak of each roof and in the shafts between the pavilions brought light to the lower galleries. We organized the construction around a new internal street in the museum, with two new entrances, north and south. The pavilions flanked the street on one side; on the other side, the street opened to a glazed courtyard-atrium that united the existing museum buildings. To give the new complex a unique and strong identity, we enclosed the street and atrium with curved glass-and-steel structures, creating a sinuous silhouette in the skyline. Beneath the curvy glass roofs, sail-like sunshades helped modulate the entering light, while evoking the great sailing ships integral to the museum's history.

Colonial-era tombstones in Salem, Massachusetts, inspired pavilion silhouettes at the Peabody Essex Museum.

As we assembled the model for our final presentation, I was troubled by the monotony of the way the end walls of the five pavilions faced the street—as identical silhouettes, side by side. I toyed with the idea of cladding them with different materials—copper, zinc, brick, wood—like treasure boxes coming off ships, but the idea seemed too capricious, too decorative. Then I recalled the beautiful variety of the seventeenth- and eighteenth-century tombstones in the

Charter Street Cemetery, which along with the Salem Witch Trials Memorial abuts the museum. Some were a simple arch, others a double concave curve, others angular and pointed, with many variations in between. Applying the shapes of the tombstones to the end walls of the galleries created a kind of musical skyline, a symphonic assembly, richer than the staccato repetition of one roof form.

I was thrilled when I learned that our design had been selected, a decision ratified in a meeting with Ned Johnson, the founder and chairman of Fidelity Investments and the principal financial underwriter of the Peabody Essex project. My staff and I were on the way to realizing the first major institutional project in our home territory of Massachusetts. But the course did not run smooth. The plan, as it continued to develop, needed to be contained within one block of a Salem city street, which required the demolition of too many of the museum's old buildings. The cost benefits were questioned. But there was a way out: incorporating a parcel on the adjacent block and making the city street in between an interior part of the museum. We suggested that a new outdoor pedestrian bypass, on museum land, could replace the street. Salem drew the line: this was not acceptable. Frustrated, Johnson and the museum board decided to move the entire museum to Boston; one place considered was a site on an abandoned bridge spanning the Fort Point Channel. My team got to work on a new design, starting from scratch. The possibilities were exciting. But in the end, they were never realized; the city of Salem wisely decided to back down.

The last surprise came from Ned Johnson himself. He was a collector of Chinese art, and returned from one of his trips to China to announce that he had bought an entire two-story eighteenth-century house in the Huizhou region and was having it disassembled and shipped to Salem. The house and courtyard, and all the furnishings, now needed to be incorporated into the design. We came up with a way to ensure that the house retained its own identity and did not undermine the larger museum plan—and yet was fully integrated into the complex. The Chinese house is one of the Peabody Essex Museum's great attractions.

* * *

xploration Place, a private, nonprofit science museum in Wichita, Kansas, was a project for which we were selected by an interview process rather than a competition and which, at a personal level, became transformative. Here, the power of place involved water. The site of the planned science museum lay along the banks of the Little Arkansas River, overlooking downtown Wichita. The contenders had been winnowed down to a short list, including Frank Gehry, Norman Foster, Hugh Hardy, and myself. In other words, I faced some very tough competition.

When I came for the interview with the selection committee—which included a representative from the Boeing Company, a major donor—I was able to walk the actual site for the first time. To say that the parcel was "on" the banks of the Little Arkansas River turned out to be misleading, because between the parcel and the river was a four-lane parkway. When I stood on the site, given the wide roadway and the narrow river, I could not see the water. But when I crossed the highway, the presence of the river became a dynamic reality. A bend in the channel created a stretch of rapids that seemed to course toward me. I came to think of the place where I was standing—looking upriver toward the rapids—as the "magic spot."

I went into the interview an hour later, described my experience, and then insisted that this was the magic spot. But the site was cut off from the magic spot by the highway. Perhaps, I suggested, an effort could be made to convince the city to realign the highway so that building directly on the river would become possible.

The committee members were taken aback but intrigued. I don't know that they knew for sure whether they could get the highway rerouted. They said when I was selected for the job only that they would make a serious effort. During the six months that followed, I began developing schemes showing what would happen if we could reroute the parkway and what the alternative would be if we could not. The mayor launched a series of public hearings. Not surprisingly, differences of opinion existed among the citizens of Wichita. Soon the Friends of McLean Boulevard was organized. For some people driving to work every morning, that brief encounter with the river was the brightest moment of the day. Runners and cyclists, for their part, thought it would be wonderful to

give the public direct access to the waterfront. In the end, the city decided to reroute the highway.

As we continued to develop the plan, I looked for ways to maximize the sense of the river's presence. At one point, I decided to try to create an island by cutting a channel and letting the river flow on both sides. The galleries would be built on this island, while the entrance, lobby, theater, and restaurants would be built on the mainland. Bridges would connect the two. In the island building, visitors would experience the river flowing against the prow and around the building, giving the illusion that the building was in motion. I had a flashback once again to Château de Chenonceau—and to my design for the Canadian Museum of Civilization on the Ottawa River.

When I presented this idea to the committee, there was a chuckle of ironic recollection, and someone spoke up, telling us that the land we were turning into an island actually used to be an island—Ackerman Island—and that the channel we proposed digging had been filled in as part of a work program during the Great Depression. It was one of those reaffirming moments. Cutting a channel and inducing a river to flow around an island building would raise complex issues of hydraulics, silting, and waterproofing. That said, the active space at Exploration Place would be designed to rest on foundations that rise above the waterline—just as Château de Chenonceau does—so direct flooding would not be a concern.

The eighteenth-century instruments at the observatory in Jaipur, India, take geometric forms.

What we would need to guard against was moisture seeping upward, a more manageable task.

There is a particular quality of the Great Plains—the flatness of the prairies, the enormous arc of sky—that kindled an obsession with the silhouette of the buildings. For several weeks I conceived of the island as a barge in the river, loaded with pavilions, each a unique shape. But that seemed too arbitrary. After all, this was a science museum. I wondered whether a mathematical mechanism could help generate the roofline of the buildings. A structure in Jaipur, India—an astronomical observatory built in 1734 by Jawai Jai Singh that has long been admired by architects—offers an example. The rich geometric forms, generated by instrumental mathematical considerations, are at once intriguing and inspiring.

What if we first developed the floorplan, assembling all the required spaces, and then imposed a roof geometry? It took many weeks of trial and error, but we came to see the possibilities inherent in the geometry of a toroid—the geometry of the surface of a donut or bagel.

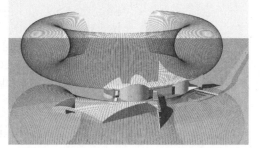

The challenge was a complicated one. Looking today at the island from above, the walls of the museum can be seen as having different configurations—sometimes they curve outward, in an arc, and sometimes they jut out as triangles. Imagine then that the roof was defined by the arc of a giant toroid. Imagine that the building, with its complex outline, is made of a soft clay, and then the larger-scale shape of that toroid is pressed into the roof surface. The center would be pressed low, and the wall edges would rise in a complex silhouette against the convex surface of the giant donut. And because each line wrapping the toroid is always the same radius, one ends up with a totally rational structure, with beams of equal curvature roofing the building.

The roofline at Exploration Place was generated mathematically; imagine a toroid pressed into the floor plan.

Moving off the island to the part of the museum located on the mainland, we reversed the process: imagine the center of the toroid being deep in the earth. The roof is shaped then by the convex curve inside the face of the toroid. Instead of a bowl-like roof facing the sky—as the island structure has—the off-island structure has a dome-like roof.

The island and landside buildings, convex and concave, then become complementary, yin and yang forms. When I think about Wichita and other projects that rely heavily on mathematical relationships, the words of the British mathematician G. H. Hardy come to mind: "The mathematician's patterns, like the painter's or the poet's, must be beautiful; the ideas, like the colors or the words, must fit together in a harmonious way. Beauty is the first test. There is no permanent place in the world for ugly mathematics."

The astonishing thing was that we were not "designing" these extraordinarily rich forms. We were deriving them from a collision between two geometries. That was quite a payoff, not just in terms of aesthetics but also in terms of efficiency. Because every beam that formed the roof had an identical curvature, the manufacture of the beams and connections could be standardized. In other words, as the geometry of the building became richer and more complex, the construction became simplified and less costly.

The term "inherently buildable" defines the Wichita experience. Complexity doesn't have to make building harder. Exploration Place turned out to be one of the most economical buildings we've ever done. The cost of the building at the time was about $130 a square foot, perhaps half the cost per foot of comparable museums.

Just as Exploration Place was winding down, another project—for a major performance space in Kansas City—was entering the planning stage. There's a venerable body of lore having to do with famous ideas being sketched out originally on napkins—like Milton Glaser's "I Love New York" campaign. The first rough rendering of the Kauffman Center for the Performing Arts really was presented by me initially on a napkin—at a dinner in 2000 with Julia Kauffman, who

Perfecting the Kauffman Center's acoustics required a precise scale model of the concert hall for testing.

played a pivotal role in the project not only as its patron but through sheer force of will. The idea was to build two exuberant shell-like structures, one for orchestral music and one for opera and ballet. If the challenge in Wichita had been to activate the river, the challenge in Kansas City was to activate the acoustics.

I was determined to create a concert hall that transcended the traditional shoebox configuration, as fine as many such venues may be. I was impressed, as many architects of my generation also were, with Hans Scharoun's Berlin Philharmonic concert hall, in Germany. There, for the first time, the public was able to surround the performers from all directions, and the concert hall became more of a happening in the round than most concert halls permit. Frank Gehry had incorporated Scharoun's insight into his Walt Disney Concert Hall, in Los Angeles, which opened in 2003. But the renowned acoustician hired by Kauffman, Russell Johnson, insisted on a shoebox configuration for orchestral music, which effectively ruled out all other ambitions.

The Kauffman Center for the Performing Arts, in Kansas City, Missouri, completed in 2011.

I remember talking with Frank Gehry one day and lamenting how difficult it had been working with Johnson. "Moshe, you are crazy," Frank responded. "You should work with Yasu Toyota. He is amazing. I could not have done Disney without him." Gehry proved to be right. Ultimately, Julia agreed to switch acousticians, and before long I was traveling with Yasu to rehearsals and public performances in concert halls, some of which he had designed, throughout Japan. Under his tutelage, I learned to hear and distinguish sounds in a new way.

Working with Yasu Toyota was a tango. One step: the architect's wishes. Second step: the acoustician's needs. Back and forth. For a concert hall, success means an ambience in which the sound is sublime everywhere, not just in some select seats. And it must be sublime for a delicate sonata, a string quartet, a full symphony, a chorus with a hundred musicians, and from time to time a rock concert. The sight lines, too, must be perfect for every seat in the house. Acoustics are about transmitting, amplifying, and mixing sounds. The shape of the room and the texture and character of every element come into play—the upholstery on the seats, the carpeting on the floor, the nature of the hard surfaces, the number of people in the chamber. Today, computer modeling such as that done by Arup SoundLab can simulate concert-hall (and other) environments, whether existing or proposed—a great advance in a field that was as much art as science.

Perhaps the greatest significance of Yasu's involvement with Kauffman was that he made it possible for the main orchestra floor—the one fronting on the stage, directly behind the conductor—to be raked to a much steeper slope. The conventional, shoebox approach requires that this area be relatively flat, which greatly compromises the experience—as one is only too aware in Boston's Symphony Hall, for instance, where the audience sits very low in relation to the stage. But by raking the orchestra floor and raising the wing walls, a feeling of true proximity was created. The audience feels drawn into the orchestra platform, and not just toward the conductor and strings at the edge of the stage but toward the entire orchestra, which is now visible: the winds, the bassoons, the oboes, the percussion.

* * *

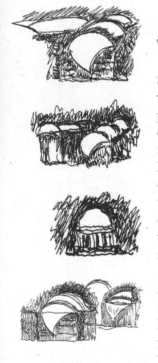

When I received a call in 2001 inviting us to participate in a competition for the headquarters of the United States Institute of Peace (USIP), my interest and curiosity were immediately aroused. I must confess that, until the phone call, I had never heard of the institute. I soon discovered that my ignorance was widely shared by colleagues and friends. When I casually asked small groups who among them knew anything about the USIP, I rarely found anyone who did. Only after visiting the group's headquarters—located at the time on the twelfth floor of an anonymous office building on Seventeenth Street, in Washington, D.C.—did I come to understand the institute's wide-ranging activities, including missions in many of the world's conflict zones.

The anonymity of existence in a downtown office block, alongside lobbying and law firms, was probably one reason for the USIP's low profile. But by good fortune, the U.S. Navy had agreed to donate to the institute a site it had been using as a parking lot. It was a strategic site—some called it "the last site on the National Mall." The tract was adjacent to the State Department, directly on Constitution Avenue, overlooking the Lincoln Memorial, and was a stone's throw away from the Vietnam War Memorial.

I realized that this proposed new headquarters had to be a building that does what other buildings on the Mall do: have a life of its own and robustly represent its specific

Notebook sketches, 2002, pertaining to the United States Institute of Peace. Jefferson Memorial, second from bottom.

The United States Institute of Peace, on the National Mall, in Washington, D.C., which opened in 2011.

mission, in this case the pursuit of peace. Every project has its own rhythm of creation and poses its own particular set of issues. The United States Institute of Peace was an invited competition. Five firms had been short-listed from among a larger group of applicants.

Basically, the USIP building was to house two activities: first, a research center or think tank, including support spaces such as a library, auditoriums, and meeting rooms, where researchers and fellows from around the world would explore the issues of peacemaking and conflict resolution; and second, a program of public outreach. There would be a museum-like, twenty-thousand-square-foot interactive space presenting narratives of the great peacemakers in history as well as the activities of the institute. Associated with that would be a large public space, a great hall, where lectures, dinners, celebrations, and other events would occur. The site is bounded by the National Mall to the south and Twenty-Third Street to the east, with the Kennedy Center for the Performing Arts and the Watergate complex visible to the west. As usual, we made a model of the terrain.

As my early sketches show, we quickly developed a scheme in which there were two atria within the building. The larger one faced the Lincoln Memorial and the Mall. A secondary, more intimate and private atrium faced the Potomac River as it bent around a curve in the direction of the Kennedy Center. The small atrium would be enclosed by four levels of offices and meeting rooms, plus the library. On the ground floor of this atrium, institute staff could meet informally and find lunch and refreshments. The larger public atrium stepped down, overlooking a great hall, which extended to an exterior terrace giving onto the Mall. With this double-atrium arrangement, almost every work space inside the building had both daylight and a view to the city beyond. The two atria opened up all interior spaces to the outside. It also did the opposite: from the exterior, those on foot or driving by would get dramatic views into the building. The life within would become visible.

But the new headquarters was not just about the efficient accommodation of program needs. By virtue of its location, the building would become a symbol of peace—the only one on the Mall, which holds more than a few commemorations of war. Moreover, the surrounding architectural context had to be considered, first the grand, classical Lincoln Memorial, inescapable across the grass; and then, beginning with the National Gallery, the line of institutional buildings along Constitution Avenue: a parade of limestone-clad variations on the classical style. The gray pile of the Depression-era State Department building took up two full blocks to the north.

I decided that the exterior of the building should echo the classical character of the institutions along the Mall. It would be built with precast concrete of limestone color, with orderly openings of varying scales, echoing the masonry structures so familiar in Washington. In contrast, the plan's interior geometry was anything but classical. The two atria were defined by curved walls and the large windows of the offices that looked into them. As the design evolved, Richard Solomon, president of the institute, became an active partner in the process. He appreciated the subtle issues involved in creating an effective and interactive workplace, while also pushing for the building to be a significant structure in the nation's capital.

I often describe the architectural design process as if there are easily distinguishable separate steps, each resolved on its own at a particular moment and in orderly fashion. But the process is not like that. Everything affects everything else. In this case the site and its topography posed the challenge of how the USIP would fit among other monumental structures, notably the Lincoln Memorial and, farther away but quite visible, the Jefferson Memorial. The question of how the institute building would manifest itself in the National Mall skyline and what would make it memorable preoccupied me from day one. We wanted a feature of the building to send an unmistakable message: this is the U.S. Institute of Peace.

I have always loved the Jefferson Memorial for its simplicity and purity—that gleaming white dome set upon the circular Ionic colonnade. I knew that the roof of the institute's atrium must be transparent or translucent; we depended on light coming through the roof to provide daylight for the working spaces. The Jefferson Memorial steered me to search for a new kind of transparency—a soft, constant glow of daylight that might then, at night, create dome-like white surfaces glowing from within. The best analogy I can think of is the two-way translucence of an eggshell. Working with industrial partners, and eventually selecting the Seele company, in Germany, we developed an eight-inch-thick shell structure made up of white glass with a fritted ceramic surface and an inner film that would modulate and diffuse the light.

Generating a geometry derived from the shapes of the atria led to a series of spherical and toroidal surfaces interpenetrating one another. From certain vantage points, the upper reaches of the building recall a bird in flight—a white dove, wings outstretched. From others, the upper reaches appear serene: whiteness, a lightness of being, a dynamic sense of motion.

*　*　*

The process of coming to understand the uniqueness and specificity of each new project—of its program, its place, and its client—is a great joy. Skirball is about the Santa Monica Mountains and fitting into the nooks and crannies of that topography. Exploration Place is about harnessing a river and the rugged vitality of water in a prairie-flat terrain. The first of these answers to clients who are an archetype of liberal Angeleno sophisticates. The second answers to clients who are no less sophisticated, in a different fashion, but have the openness and pragmatism of the Midwest and the Great Plains. An architect needs sensitive antennae to pick up qualities of difference.

In the decade since the opening of Crystal Bridges, I have stayed in close touch with the institution. It has become the center of community life in the Bentonville region. Attendance, which was three hundred thousand visitors a year at the time of the opening, had expanded to seven hundred thousand visitors.

On a visit to Crystal Bridges in 2018, on the occasion of a concert in the Great Hall by Yo-Yo Ma and his Silkroad Ensemble, I walked with Alice around the site. She was thrilled by how well the project had been received by the community and asked me to create a twenty-five-year master plan. We imagined new galleries and curatorial-support facilities. We looked for ways to increase capacity by expanding the lobbies and parking and enhancing access. We also explored the potential of the rest of the Crystal Bridges property. We did not limit our effort to master planning, but developed building designs in full, and presented the design with a model of the entire site, showing both the existing buildings and those proposed. As Alice and the board reviewed our submission, there was increasing enthusiasm, and a decision was made to move ahead quickly. As Alice put it, why wait twenty-five years? A new wing of the museum is already taking shape, in the form of a new bridge, this one with window-lined galleries where glass, metals, ceramics, and other objects that are less sensitive to light will be shown in a way that comes as close to being outdoors as the indoors can offer.

Crystal Bridges, since its opening, has paid special attention to architecture in addition to American art. Several years ago, Alice purchased Frank Lloyd Wright's 1954 Bachman-Wilson House, in New Jersey—one of his so-called Usonian houses, single-story homes designed for middle-income families. She had it taken apart and reassembled on the grounds of the museum. She also acquired and has put on display a Buckminster Fuller dome. Given a growing need for housing—for visiting artists and others—we incorporated a residential cluster of some two hundred apartment units into the master plan. It was to be a model community built with advanced timber technology—prefabricated, modular—a Timber Habitat for Arkansas, with gardens and terraces for all units. Local contractors have been invited to submit proposals. In the context of the Timber Habitat, we are also thinking about reproducing one of the Habitat '67 apartments in Montreal as an exhibit, so that it can be experienced together with Wright's Usonian house—showcasing important aspects of residential development during the past century.

In 2020, I got a call from Alice, telling me that she had decided to build a home for herself on Beaver Lake. I had not traveled since the start of the pandemic, but I knew it was impossible to design a house without seeing the site. Alice had sent videos of the lake, with a hill rising to the south, but it was not enough—I could not "feel" the setting. Alice sent her plane to get me, so that I could avoid commercial flights.

The next day, we drove through the yellow pine forest to Beaver Lake and walked the land. The sunset was extraordinarily beautiful. The location considered by Alice, up on the hill, had seemed promising to me before I experienced the setting firsthand. But I changed my mind: placing the house downhill, directly by the water—bringing the lake, so to speak, into the home—would be magical. Alice concurred, even deciding to name the new home Convergence. A house on a hill may be dramatic, but a house on the shore would be intimate. By the time of my departure, the following day, I had a version of the house plan sketched out.

The power of place.

Raffles City Chongqing, in the People's Republic of China.

CHAPTER NINE

Megascale

M arina Bay Sands, a so-called integrated resort, is a waterfront complex that has become Singapore's national landmark. It is the background image in daily newscasts and has been showcased in movies such as *Crazy Rich Asians*. It features hotels, shopping malls, pools, casinos, museums, theaters, and a convention center. The three defining towers are linked at the top by a park in the sky.

At the time of its commission, in 2005, Marina Bay Sands was by far the most complex project our office had undertaken. Its nine million square feet were designed and built in four years, at times with more than ten thousand laborers, working in three shifts, seven days a week. Its owner, the late Sheldon Adelson, the casino magnate and chairman of the development corporation Las Vegas Sands, is by far the most difficult client I have had—a man I never would have expected to seek out my services, given the great differences in our temperaments and our politics. And yet, we ended up collaborating on one of the most successful—on many different levels—buildings of the twenty-first century. And we decided to collaborate again to extend and expand the original project.

Strangely, the story of Marina Bay Sands began at Yad Vashem, in 2005. Michal and I were in Jerusalem for the dedication of the Holocaust History Museum. It was a bitterly cold night, the ceremony was outdoors, and after the speeches I had joined a long line of patrons waiting for the washrooms. Rolling up to me in a wheelchair came Adelson. Sheldon and his wife, Miriam, had been major donors to the Holocaust History Museum; indeed, they had just given it $50 million. Coming to a stop in his wheelchair, Adelson complimented me on how the museum had turned out and went on to say that he was hoping to develop a resort on the waterfront in Singapore—there had been a call for proposals by the government. He had been working on it with his Las Vegas architect, but the Singaporeans were telling him they did not want Las Vegas architecture. They wanted something in a completely different direction—something more contemporary and in the spirit of Singapore. He asked if I would be interested in the project.

This was an odd conversation to be having while waiting to use the men's room at Yad Vashem. The emotional and historic distance between Yad Vashem and an Asian gaming resort could hardly be greater. I indicated my potential interest but thought no more about it. A couple of weeks went by. Then came the call from Las Vegas: Could I come out immediately? Of the year provided to the bidders to submit proposals, four months were left. A deadline was imminent. Adelson needed a design for seven million square feet, including all the presentation material—films, renderings, models, and much more—within four months. I was aware of Adelson's very conservative and eclectic taste. His major architectural statement thus far was the Venetian, in Las Vegas, which reproduces the Saint Mark's bell tower and, on the interior, replicates the trappings of a Renaissance palazzo. I was skeptical that this sensibility had much in common with the design spirit and ambitions of contemporary Singapore.

When Adelson had first approached me, it was on his part more an act of faith than opportunism, considering that I had never designed a casino or an integrated resort or built anything on that scale. He had no idea what I would be able to produce. His overture to me was based on an appreciation of what I had done at Yad Vashem. For my part, I saw the offer as an opportunity to undertake a project of great architectural significance. What interested me was not the casino but the chance to create a quintessential mixed-use complex that provides an exemplary public realm.

Revitalizing the public realm is among the great challenges of our time. Say the word "city" and one immediately conjures those places where we come together as a community, as a society. If a city is akin to a house, the public realm is its living room. The living rooms of a city are the agoras and forums and souks; the boulevards and gallerias; the city squares and city parks—urban spaces that have evolved in every culture. The public realm has always facilitated, and often celebrated, the full range of human engagement. It includes places of governance—palaces, parliaments, courthouses—and the public spaces around them. It includes places of commerce—bazaars and markets, harbors and railway stations. And at its best, the public realm includes places of civic life—the piazzas and gardens and other formal and informal meeting places where we gather and interact and enjoy life beyond our homes and our workplaces.

The public realm always mirrors the society that creates it. Societies in which there are many rituals—military parades, large-scale communal worship—tend to create public venues to match: think of the enormous expanse of Red Square, in Moscow, or of the Zócalo, in Mexico City; or, by contrast, the simple common of New England towns, where communal life is more low-key. The public realm has a wide range of scales and purposes—from the vast Piazza del Duomo, focused on the cathedral, in Milan; to the more-intimate Piazza della Scala, in the same city, focused on the opera; to the two piazzas linked by the glazed Galleria Vittorio Emanuele II, also in Milan, focused on commerce.

The cities of the twentieth and twenty-first centuries have faced immense disruptions that have undermined or destroyed centuries of accumulated tradition. The automobile not only clogged streets everywhere but also encouraged cities to expand like untended vines. Crisscrossing highways effectively turned neighborhoods into cages. Elevators and air-conditioning, meanwhile, have allowed buildings to expand in size, vertically and horizontally, to an almost unlimited degree, creating unprecedented urban density without careful thought of consequences. The great shopping streets, where carriages and people once mingled in shared space, have given way to cars, parking, and ill-conceived infrastructure—sometimes with the philosophical blessing of prominent voices. "The street wears us out," Le Corbusier complained in an article in 1929. "It is altogether disgusting. Why, then, does it still exist?"

The outcome of all this has been the increasing privatization of the public realm. Some historic cities have been able to maintain the richness of their heritage and their public character—cities such as Paris, Amsterdam, and even New York. The epitome of privatization is the shopping mall, a space dedicated essentially to a single use. Only those who will spend and sustain the mall's economy are truly welcome. Most often, this new kind of privatized public realm is not organically connected to the street level. And it is part of the strategy of mall designers to lure people in and then, literally, to help them get lost so they can't find their way out.

To be sure, malls possess attractive novelties (easy parking, air conditioning). But after several decades of the mall as a dominant typology, there is a growing awareness of its shortcomings. People crave contact with fresh air and nature. People also want access to a more diverse range of offerings than just shopping. They want entertainment and culture, libraries, civic facilities—all of which also bring everyone into contact with a greater variety of human life and social backgrounds. People want a sense that they are part of a larger communal identity.

This became obvious and vividly felt during the coronavirus pandemic, as people fantasized about being back in the public realm. People want a sense that a city is not a vehicle of isolation and entrapment but rather a living entity that "contains multitudes," in Walt Whitman's phrase. They want a place that both embodies and makes possible the highest human aspirations. The opportunity to create such a place was, for me, the great appeal of a project such as Marina Bay Sands.

I told Adelson when he called that I'd have to mobilize my entire office, and that the cost for the submission would be approximately $2 million—but that I believed we could do it. A day after our conversation, Adelson said, "Go for it." The schedule was tight, and we dove right in.

* * *

It's easy to write the words "mobilize my entire office" but not so easy to actually do it. A firm such as ours, with four or five major projects going at once, yet with a relatively small staff, must operate like a finely calibrated clockwork mechanism. I am responsible for the central ideas and designs, and I stay involved at every stage of every project, but only a skilled professional team can keep concepts evolving and ensure that the mechanism functions at a superior level.

Imagining a recent day at the Somerville office provides a snapshot of the internal orchestration required. On a typical morning, I might leave my personal office, make a sharp right, and enter an airy, open space to visit a team that is working on a preliminary design for a project in its early stages—the new integrated resort we have been commissioned to build in Tokyo Bay. The team members have massed the program in blocks of foam, the foam itself painted different colors to represent different functions. They have also made a model of the site. In my notebook, I have a variety of sketches with a range of alternative schemes. We sit and talk about where to place the major masses; how to relate the masses to the view; how to orient the building with respect to the sun, the wind, the terrain. This is the moment when one begins to understand how a building is going to be organized, before one gives it form. It is also the moment when one identifies the distinct kinds of expertise that the project will need: the various types of engineer, the landscape architects, the lighting consultants, and others.

Next, on the opposite side of the office, I spend time with another team, this one working on a project that is much further along. There may be two or three study models—some of them in foam, some of them models in wood, which is far more precise. There are already trees stuck in certain places. On this particular day, I am looking at the new Facebook campus, where the overall idea is already established, but we're trying to figure out the configuration of the offices and how exactly they relate to the landscape areas surrounding them. At the same time, we're looking at code issues, because we've gotten word that we need more fire stairs than we thought. We've also just received a memo from the mechanical engineers listing all the spaces they need for ducts. We are studying ways to screen the messy tangle while providing the necessary access to service the mechanical systems.

In another part of the office, I join the team that is creating the detailed plans, the working drawings, for the Crystal Bridges addition. Here we're looking at highly detailed drawings. The drawings are already on the computer, and we're looking at screens. I find the screens frustrating, because I like to draw and scribble over details as we refine them. In the old days we had big sets of drawings in the office, and we'd reprint them every two or three weeks as the job progressed. We would go through the sheets and "red-mark" them, turning the large sheets one at a time. This was when we really got down into the weeds.

For the Crystal Bridges addition, we conceived the major new gallery for changing exhibitions to be lit by north-facing skylights—traditionally assumed to be superior because they avoid direct sunlight. In the process of detailing this with our lighting consultants, we built a large model, which we studied outdoors. To our surprise—verified by computer simulation—the daylight within the model was very "cool."

The reds and yellows were dull—a no-no for an art gallery. The team has been drawing options to address the issue, such as attaching an exterior reflective metal fin in order to reflect sunlight into the skylight, blending it with the cooler light. The adjusted model confirmed that this blending would produce light that was balanced across the spectrum. Now, as we examine the large-scale model, we're considering how to integrate the fin's attachment to the roof, and also how to incorporate sprinklers, track lights, air-conditioning grills, speakers, and laser-beam smoke detectors into the gallery's interior. It's a handful. All told, we've been studying this for over two months.

A typical day doesn't go by without my going to the model shop two or three times. Sometimes it is to help resolve minute details or select colors and textures for a presentation to a client. A model that will be transported to the client is internally lit and can be photographed to simulate various times of day and night. We spend a long time discussing the appropriate trees, which we make by hand in the office, one by one. Every detail can have an impact. How can we get ponds in the Crystal Bridges model, for example, to both have a little ripple and also be reflective? Over the years we've developed techniques: we apply one layer of rippled acrylic and another layer of smooth acrylic and a final layer of blue color. You see the ripple when you look vertically, and you see the reflection when you look horizontally.

After some time in the model shop, I may stop by the materials and interior department, with its shelves and shelves of samples: stone, glass, metals, fabrics, woods, tiles, laminates. For every project, we keep a drawer that holds an example of each type of material used.

About 25 percent of my time on a typical day is devoted to management. I may have a meeting about staffing because a very large project that was just about to start has been delayed by four months. So, all of a sudden, we have seven or eight people who need reassignment—assuming there's something to assign them to. (This is a perpetual problem in an architect's office, with its unexpected lulls and peaks.) Invariably there will be a meeting about finance—where various projects stand financially—and often there will be one about legal matters, including the inevitable nuisance lawsuits. For instance, someone slips and falls in a piazza of one of our projects, and the injured person decides to sue not just the owner of the building but the architect. Finally, I'll usually spend part of the day in the office library, helping to prepare formal presentations to prospective clients—in other words, generating new business—or preparing for a coming lecture.

To understand a typical day at our office is to understand something else: what it requires to suddenly pivot away from ongoing projects and pivot toward something else. Mobilizing the office for Marina Bay Sands meant putting a typical day on pause—for months. It meant taking, all told, thirty or forty people away from what they were doing and putting them on an emergency footing. It also required maneuvering and diplomacy to avoid antagonizing existing clients. We did all this—successfully—and we did it willingly because the stakes were high.

* * *

I have great admiration for Singapore. The small island nation at the tip of the Malay Peninsula has its critics— they find its politics overbearing and authoritarian, its culture rigid and antiseptic—but the country's achievements have been nothing short of astonishing. Singapore has been bound up with my life for more than four decades. I hold Israeli, Canadian, and U.S. passports, and I often say that by now I should have Singaporean citizenship too. Sheldon Adelson had no idea about this when he rolled up to me at Yad Vashem.

In the 1960s, '70s, and '80s, Singapore was transforming its society. The country, a former British colony that became independent in 1965, had the good fortune to find itself in the hands of a benevolent founding father, Lee Kuan Yew, a visionary leader who understood that the small city-state would need to evolve its own unique institutions and systems of governance. It would be governed by the rule of law, but it did not embrace the full range of rights and protections that characterize most Western democracies. Singapore preserved elements of the British legal system. The country was not corrupt. It was scrupulous about freedom of religion and equal opportunity for its minorities. The press, though, was well controlled. Dissidents had a hard time. Singapore was, and is, essentially a one-party state.

But Lee believed in planning, education, housing, and health care for all. Economically, he was assisted by a brilliant group of technocrats who helped leverage Singapore into its current position as a financial center, regional entrepôt, and platform for specialized manufacturing. It is a wealthy country. Lee was very much a social engineer. Every housing project was demographically mixed; none were purely Chinese or Malay. The government famously instituted fines for chewing gum and spitting on the streets. Painting graffiti could get one caned. Dealing drugs was punishable by death. Singapore today is the cleanest city in the world. It is also among the greenest. Every major street became a tree-lined boulevard. Every bridge seems to be covered with flowering plants. City planning is interventionist and forward looking. Singapore conceives infrastructure a generation ahead of demand. It builds new airport terminals before it needs them. The same can be said of highways, subways, and landfills. Civil servants are highly educated and well paid.

My first visit to Singapore had been scheduled for 1973, when I was returning from China after the Canadian state visit with Pierre Trudeau. That had been the plan. The outbreak of the Yom Kippur War intervened, causing a sudden change. Then, in 1977, out of the blue, Robin Loh had arrived in Montreal and taken me to dinner at the Ritz-Carlton. We worked together on a number of projects in Singapore, including Ardmore Habitat, which opened in the mid-1980s, and over time I developed my own network there. In the late-1990s, we started work on a triple-tower luxury-apartment complex at Cairnhill Road known as The Edge, in which each apartment occupied a whole floor of the building, and the three buildings were connected by bridges at each level. Over time, I began developing relationships with Singapore's Urban Redevelopment Authority and Housing and Development Board. These are crucial relationships to have. The URA is responsible for planning and urban design in Singapore, and the HDB builds and provides publicly assisted housing for 80 percent of all Singaporeans.

Bygone era: the port at Singapore, c. 1930.

In the mid-1980s, representatives of the HDB approached me with a proposal to undertake an initiative they were launching to explore fresh ideas for their new towns, many of which they felt were monotonous and devoid of character. They suggested that I undertake a project and link it to the Graduate School of Design—drawing on and collaborating with Singaporean students in the Harvard program. (One of these was Fun Siew Leng, today the chief urban designer at the URA.) We worked in Boston for a few months on a plan for a new waterfront town of 125,000, called Simpang, then traveled to Singapore and presented it to the HDB.

With the developer Sheldon Adelson and his wife, Miriam, viewing a model of Marina Bay Sands, Singapore, 2008.

The central innovation was that, instead of a single building typology—essentially, repetitive, twenty-story slab-like buildings that faced each other in a confrontational way, which was the prevailing model at the time—we introduced several typologies, both high-rise and mid-rise. The mid-rise buildings created breathing space between the taller towers. There was also a waterfront building type, towers with apartments bridging across the top, like giant aqueducts. The waterfront buildings were designed to create large "urban windows" that gave buildings behind them unobstructed views of the sea. The whole plan had an airiness to it—it didn't feel congested or dense. Everyone had ample daylight, open views, and access to green space.

The first reaction at the presentation was one of disbelief. The HDB felt these effects were achieved at the expense of the densities it had required. I replied that we had indeed achieved the necessary densities. It took a couple of days to dispel the skepticism—two young architects were assigned the task of measuring, counting, and confirming that the densities were on target. In the end, given the mindset of that moment, the plan proved too unconventional and adventurous for HDB. It was shelved for later.

Fast-forward to Sheldon Adelson. He was a deeply opinionated man. His capacity for grasping the big picture—an uncanny understanding of the larger forces at play—did not diminish his compulsive addiction to the little picture. I have communications in which his peremptory voice, though muffled and softened by the diplomatic language of intermediaries, still comes through. Here is an email from one

of Adelson's associates to me, recounting a conversation he
had just had: "His current rage was the hotel rooms, which
he 'hated,' asking 'who designed them?'" Adelson once sent
a memo about escalators, insisting for some reason that they
be parallel to the gaming tables at the casino. Later, when we
were just about to open Marina Bay Sands, Adelson arrived
and took a tape measure and started measuring the distance
between the casino tables. He had a theory that there is an
optimal amount of distance between tables, no more, no less,
and if you have three inches extra, you're wasting space. He
also counted the number of toilet stalls and measured the
distance between them.

I will not soon forget spending a fifteen-hour stretch
with Adelson on his private 747, taking him through the
plans for Marina Bay Sands. (He eventually upgraded to an
Airbus, capable of making the Las Vegas–Singapore trip
nonstop.) I had come to realize that Adelson was not always
fully informed about the progress of the design, and that this
could backfire, so it was crucial to spend much of the flight
reviewing the details. There were four of us: Sheldon; his
wife, Miriam; Rob Goldstein, then the vice president (and
now the chairman and chief executive officer) of Sheldon's
company, Las Vegas Sands; and me. The aircraft included
a salon, a dining room, and a master bedroom suite. Rob
and I shared the children's bedroom. Some twenty security
guards, assistants, and chefs occupied the rear section. The
scene was somewhat surreal. The 747 had previously been
used for luxury service in Bahrain and Brunei, and it still
had the original decor. Everything was gilded, down to the
faucets in the bathrooms.

I am not unaware of the paradox of my relationship
with Sheldon. Yet he enabled me to design one of the most
significant projects of my career and one that turned out to
be iconic. It is also the most profitable architectural project ever built, with pre-pandemic revenues (now bouncing
back) of $100 million per month, maintained year in and
year out—generated not only by the casino but by the hotel,
the shops, and the convention center. The government of
Singapore gladly shares in the success.

* * *

Marina Bay Sands was developed in two distinct phases. The first was the competition-design phase, done under great pressure in four months. Following our selection, in May 2006, came the second phase: four intense years of developing the scheme and overseeing construction. Only a cooperative, complex process of decision-making could complete a project of the quality that it ended up being. It helped that Adelson had a superb team. It was headed by Brad Stone, the executive vice president, who took over the management of the design, and included a number of extremely able architects, engineers, and construction people.

All of this does not, however, explain the outcome. The architect-developer relationship is fraught with tensions that in other circumstances would have made it impossible to realize radical ideas such as the Marina Bay Sands Sky-Park. The energizing dynamic was a three-way relationship among the government of Singapore, the Las Vegas Sands corporation, and our team. The way the Singaporeans structured the deal gave the government extraordinary influence and power when certain decisions had to be made. Because we effectively had two clients—Adelson and Singapore—it was often possible, even if difficult, to navigate and maneuver between them.

The so-called Madaba map, a sixth-century CE floor mosaic that shows the *cardo maximus* in Jerusalem.

The Singaporeans wisely set the tone at the outset. They established a fixed price for the land when calling for proposals, making it clear that the key criteria for selection would be a combination of the proposed program of facilities, the quality of the design, and the capability of the developer to deliver. Had they set it up conventionally, with the highest bidder winning, they would have had very little leverage as things went forward. The Singaporean government also established clear guidelines for urban design and other objectives, covering everything from the amount of public space required, the amount of green space, the density, the view corridors, and the degree of integration with the rest of the city. The guidelines were binding, and they were ambitious and far-reaching. Once selection took place, the government set up monthly meeting reviews in which design progress was presented. A team of architects, urban designers, engineers,

At Marina Bay
Sands, a modern
cardo was
connected to
parks and the
waterfront.

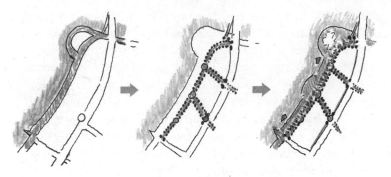

and representatives of the tourist development board were in
attendance. If there was any attempt to withdraw from some
of the commitments, they would insist on our honoring the
deal. For example, when we once floated the possibility of
reducing the cantilever—the unsupported portion—of the
SkyPark from about 210 feet to 130 feet, which would have
yielded a major saving in cost, we were turned down. It was
the strength of this three-way partnership that made the
project what it became.

I saw, in the particular mix of uses and programs, and in
the accompanying guidelines, an opportunity. The challenge
was to design an exemplary high-density, mixed-use public place connected to nature, to the waterfront, and to the
surrounding city—but with its vast size broken down into
smaller components to achieve a comfortable human scale.
Everything flowed from that objective.

The centerpiece was a new kind of spine for the project, integrating the shopping galleria and the waterfront
promenade into a singular multilevel, indoor-outdoor linear space filled with daylight and open to views of the water
and downtown Singapore. It ran parallel to the shoreline
from north to south. This served a function similar to that
of the *cardo maximus* of ancient Roman cities. But it was also
a climate-convertible space, with doors that opened at night
to the tropical air in pleasant weather. Meanwhile, two galleries would run perpendicular to the *cardo*, connecting the
complex to the bay. As in ancient cities, every component of
the program—in this case, convention center, casino, theaters,
museums, and so on—plugged into the spine. This complex
within the city was itself a microcosm of a city.

The *cardo maximus*, along with the facilities plugged into
it, formed a five-story-high podium along the waterfront.

We then located the hotel tower off the podium to a parcel to the east, away from the promenade, where it would not overpower the activities within and around the podium. It soon became clear that a single, massive hotel tower, fifty-nine stories high and more than six hundred feet wide, containing approximately three thousand rooms, would form a kind of Chinese wall, cutting off the downtown's view of the open sea. We therefore proposed to break that single tower into three, leaving generous openings between them.

We had constructed a model with the three hotel towers at our office. Models, as noted, are indispensable aids in evolving a design. We now considered another issue: Marina Bay Sands was to be an integrated resort, but the term "resort" implies an ambitious program of swimming pools, gardens, and other outdoor spaces. Where these things would go, given that the site was now occupied by three buildings rather than one, remained an outstanding question. The conventional answer (as in Las Vegas) is to put them on the roof of the podium—the portion of a complex that serves as its base and from which towers typically rise. That was not an option here for various reasons, including the fact that the podium roofing would be shadowed by the hotel towers for several hours a day. Standing by the model, with its three towers, I remember saying to myself, Well, why not? I took a long piece of balsa wood that was on the table and put it across the top of the towers. We keep buckets of lichen in the model

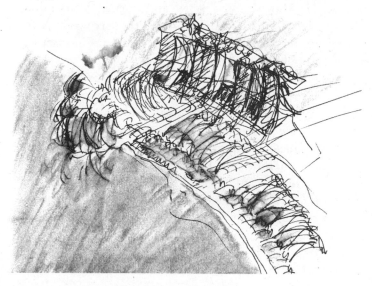

A notebook sketch from 2005 gives dimensionality to the idea.

shop that we use to make trees. I took handfuls of lichen and spread them on top, creating an instant park. I affixed a strip of blue paper, and we had an infinity pool. It made so much sense: a sky park, with unbelievable views to Indonesia, to Malaysia, to the downtown, to the harbor, to the sea.

The SkyPark idea may seem, in the telling, like a flash of inspiration, but in truth, ideas don't generate in a vacuum. I had been exploring the idea of sky parks and sky gardens—public and private—for half a century. It was fundamental to Habitat '67 and the many Habitats thereafter. The idea of expanding landscape three-dimensionally is one we had indeed been experimenting with, and building, for years.

One unavoidable step in any building process in Singapore or China is consultation with a feng shui master, a geomancer usually appointed by the client. Feng shui is not an empirical process; if you are working on a project and you consult two or three different feng shui masters, you may well get two or three different recommendations. Certain rules do, however, recur. Aggressively pointy forms are not good—negative energy. Water has redeeming qualities—positive energy. Some principles have to do with orientation and the direction of light. Feng shui is not totally capricious.

There was a time, before experience working in Asia taught me better, when my attitude toward feng shui was dismissive: this was nothing more than superstition. That attitude didn't last very long. If feng shui is important to the people who are going to live and work in the building, then it can't be ignored. The intriguing part was figuring out how to incorporate feng shui in a way that was inspiring architecturally.

At Marina Bay Sands, I had placed the SkyPark atop the three towers with the platform overhanging symmetrically on both ends. The feng shui adviser was averse to this symmetry. He had a complicated explanation—memories of the Japanese occupation during World War II came into play somehow—but his conclusion was that symmetry should be avoided. In response, we moved the SkyPark off-center, so that one end of it was almost flush with the side of the southern tower, and at the north tower it was cantilevered out to a distance of about 210 feet. Suddenly, the design became dynamic—the whole message of the building changed. There was a *whoooosh*. It was in motion—a powerful symbol of a dynamic Singapore. It was as if the park had become a skateboard.

Of course, *whoooosh* is easier to say than to build. Tall structures have to contend with the powerful force of wind. Their design has to take into account potential earthquakes. Large structures—not just skyscrapers but any substantial building, and of course bridges and roads—also expand and contract significantly with temperature changes. Given these considerations, architects break large buildings and other structures into component parts linked by expansion joints that permit the necessary flexibility. People generally don't notice them, even though they are often in plain sight. Expansion joints are what cause the rhythmic vibration of a car's tires as one drives across the joints on a suspension bridge.

At Marina Bay Sands, engineering became a major challenge. Our structural-engineering consultant, Arup Group Limited, was one of the leading engineering firms in the world—founded by Ove Arup and famous for the Sydney Opera House and, later, the Centre Pompidou. The foundations of the Singapore project lay essentially in muck—the land is low-lying, the water table high. This is not unusual; people are generally not aware that many buildings they take for granted, such as those in much of lower Manhattan, have their roots in water. To plant the towers and the podium we first had to sink giant steel-and-concrete rings into Singapore's muck. The usable basements at Marina Bay Sands were designed to sit four or five levels below the water table; the giant rings went down even deeper than that. The exterior pressures were enormous. Then, inside the rings, the muck was removed, and we built up the foundation floors, layer by layer, creating structures that could themselves withstand the pressures. The structures were built with multiple hulls, so that water seeping in could be continually pumped out. Eventually the rings themselves were removed. The pumps are permanent.

The cantilevered north end of the SkyPark extends for about 210 feet.

With large complexes such as Marina Bay Sands, coping with security and catastrophic threats becomes a prime preoccupation for the architectural and engineering teams. As we began work on Marina Bay Sands, we listed some of the issues we had to respond to with design solutions. They included but were not limited to blast engineering, chemical and biological attack, criminal attack, mail screening, electronic systems, CCTV, and intruder detection. Fire is a possibility in any structure, and the prevention of fire and the mitigation of its hazards is a profound responsibility. That requires complicated systems and networks for everything from smoke evacuation to flood control to ventilation inputs, all of which must be addressed and then approved by the relevant authorities.

The technical issues any project presents can be trivial—

Testing the expansion-joint mock-up; massive expansion joints allow the SkyPark at Marina Bay Sands to accommodate three swaying towers.

will faucets be operated by hand or by sensor?—or hugely significant. For example, each of the three hotel towers at Marina Bay Sands could sway as much as two feet in a major storm or earthquake. They would sway much more were it not for dampers, heavy elements installed high up in any tall building that are designed to begin moving counter to any oscillation that occurs, thus "dampening" the movement and restraining it. The same is true of major bridge structures. I recall the test being made by the Arup engineers of the SkyPark's cantilever. With rock music blaring, some 150 engineers were jumping up and down at the tip as others took seismic measurements. At one point it started oscillating, like a diving board. I was there at the tip, being interviewed by Martha Stewart. As the cantilever started oscillating, she threw down the microphone and took off for the main tower. (The engineering passed the test.)

Further complicating matters, the three towers at Marina Bay Sands oscillate independently of one another. The SkyPark, which connects all three towers, therefore had to be able to accommodate those movements; otherwise, we knew, it would simply break into three pieces. Making the challenge greater was the fact that the SkyPark plan called for a 500-foot-long infinity pool that spanned across the three

towers and held about 375,000 gallons of water, weighing more
than fifteen hundred tons. Someone suggested, "Just break
it up into three pools." But we wanted a single continuous
sheet of water. An engineering team devised a way in which
the various sections of the SkyPark could each move more
than eighteen inches while maintaining a watertight condition
in the pool. To date, we've never lost water.

Adelson's staff had been elated by the SkyPark concept
from the start. Adelson himself hated it. I remember the
moment when I first presented the SkyPark to him. Adelson
responded with colorful language and told me to take the
SkyPark off, calling it a stupid idea. "And anyhow," he went
on, "Asians don't swim much." (He would later tell people
that he had to persuade me to add the SkyPark, arguing that
we needed space for swimming pools.)

To break the impasse—I had threatened to walk away if
the SkyPark concept was rejected—we decided to submit two
versions to the Urban Redevelopment Authority, one with
and one without the SkyPark. In the model, we attached
the SkyPark to the towers with removable screws. But when
the time came to present it to the URA, we never got to the
point where we needed to remove it.

Adelson was convinced that his integrated resort, as
I had designed it, was doomed. In the spring of 2006, he
called the director of Yad Vashem, to whom he was close, and
complained, "I've lost this one. Safdie really let me down."
A few weeks later came a phone call from the authorities in
Singapore. The design with the SkyPark had won. So, there
was a happy ending, for the time being. Marina Bay Sands
opened in 2010, at a cost of $5.7 billion.

Adelson's lawsuit would come a few years later. In effect,
he sued me for stealing my own idea.

* * *

Years before Marina Bay Sands was a reality, while we were
working on the design submission, we prepared a film for
the government of Singapore. The selection process was highly
competitive, and we knew that all four competing developers
and their architects would be submitting packages before
meeting with the committee face-to-face. The objective of
our film was, therefore, to communicate the essence of our

design: this was more than just a building; this was a micro-cosm of a city. When the introductory filming was done, with Adelson, his CEO, and myself on camera, I felt something was missing. I requested that the camera roll again, and asked, "What about the magic? What is it that makes architecture memorable, an experience that stays with you, that moves and transforms you?" I was looking beyond functional and operational efficiencies to a larger question: Can a project transform the experience of the city? Will it draw people into the public realm? Will it become a symbol of a community—indeed, of a country?

When experiencing the greatest architectural achieve-ments, one experiences this very seamlessness, even if subliminally. An architecture or engineering degree is not a prerequisite for apprehending and appreciating the bal-ance of forces that define a medieval cathedral—the physical stresses transmitted through the stone that make possible the immaterial penetration of light. "How is it made?" and "How will it make one feel?" are not unrelated questions.

Frank Lloyd Wright's Guggen-heim Museum, which opened in New York in 1959, not long before I went to work for Kahn in Philadel-phia, is virtually a case study of that proposition. One enters directly into a singular space, softly daylit from above, the sinuous spiral lift-ing the spirit upward. The building's curved street presence foretells the space within: it is the positive of the other's negative. There is something refreshing about being in a building in which what you see is what you get. All concrete, all painted white, the canti-levered spiraling galleries creating a sense of lightness. The continual change in altitude as one climbs or descends the ramp offers ever-changing vistas of conical geometry. Looking across the central space, one sees art at a distance and also people looking at the art—producing a sense of prelude and community, and an emotional charge.

Frank Lloyd Wright's Guggenheim Museum, New York City, completed in 1959.

The Bibliothèque
Nationale, in
Paris, designed by
Henri Labrouste,
completed in
1868.

At the very moment the Guggenheim was opening, the famed Sydney Opera House, in Australia, was being designed by Jørn Utzon. The opera house provides a contrast. I felt a joyous thrill when I first saw the building, with those white, sail-like vaults, billowing upward and seemingly in motion. But unlike the Guggenheim—unlike Chartres Cathedral, for that matter—the promise of the exterior dissipates inside. The lobbies and halls and performance spaces are a disappointment, bearing little relationship to the evocative exterior volumes. The exterior is a shell, and there is no organic relationship between the exterior shell and the interior space. Utzon resigned in protest when his plans for the interior were not followed.

When I think about the marriage of technology and design in the service of functionality and transcendence, I recall a space that many people do not know: the reading room of the Bibliothèque Nationale de France, in Paris, designed by Henri Labrouste in the 1860s. Approaching the library along the rue de Richelieu, one comes upon a familiar-looking Parisian monument with its classical colonnades and its large, arched windows overlooking a pair of courtyards. Enter the reading room, and one's breath is taken away. Nine domes, each with an oculus at its center, hover over the vast space, light as a feather, as if defying gravity. Sixteen slender columns, appearing as thin as pencils, extend through the space to support them. Light flows from above and from large, arcaded windows in one of the reading room's enclosing walls. To achieve this effect—while at the same time building infrastructure capable of holding the hundreds of thousands of books that warm the walls—Labrouste incorporated advanced technology: cast-iron columns and steel domes. From these sturdy metals, exposed for all to see, he derived a structure that possesses the delicacy of fine lace. For the nineteenth-century spectator, this would have seemed an astonishing feat, refreshing and exciting. It is astonishing even now.

In the case of Marina Bay Sands, the ambitions and aspirations of Singaporeans, manifest in their actions and words, proved both humbling and inspiring. But as one designs a project and brings it to completion, the reception remains an unknown. Ultimately, the public decides what counts as a success.

Twelve years after the completion of Marina Bay Sands, it is clear that the magic was and is there. Within a year of its opening in 2010, the silhouette of Marina Bay Sands had effectively replaced the image of the Merlion—a mythic creature with the head of a lion and the body of a fish—as the symbol of Singapore.

There is no simple answer for how to create magic in architecture. And there is no single form of magic. What one might create for a place of entertainment in the heart of a city must be different from the magic one might create for a place of worship, a national memorial, a public library, or a performing-arts center. But we know magic when we see it.

The residential complex Sky Habitat, in Singapore, completed in 2015.

There was a time when I voiced skepticism that architects could ever achieve the kind of emotional engagement we experience when listening to great music. How could architecture hope to match music's many moods: celebration, happiness, mourning, sorrow? Architecture is in fact capable of these things. Each of us has known moments when architecture has conjured the transcendent. But I do believe that we—meaning society as a whole but also many architects—have lost a sense of ambition as to what we can aspire to achieve when it comes to the built environment.

We need to insist on magic. And we need to remember that, in making it happen, money plays a role, but vision and will are the essential ingredients.

* * *

Marina Bay Sands established our firm in Singapore as something of a local hero. Several things happened in the aftermath. One of them comes under the rubric of what might be called "internal affairs"—reorganizing the office. The completion of a project of this magnitude placed us in new territory. There was now no project too large or complex to take on. The office had evolved into a formidable team. That said, inside any organization, the combination of great success and great competence can result, over time, in a loss of momentum. There may be less willingness to take risks. Valued younger members of the staff may have no way to move up—and therefore take their gifts elsewhere. In 2012, in a moment of decisiveness that was unusual when it came to personnel matters, I reorganized Safdie Architects. A number of veterans retired. A small group of younger architects—Charu Kokate, Jaron Lubin, Christopher Mulvey, Sean Scensor , and Carrie Yoo—obtained equity in the firm, and I expanded their responsibilities. Restructuring of this kind is never easy. It is as much about the functioning of a creative system as it is about the talents of individuals.

A rendering of the three bridges that link the Sky Habitat buildings.

It was also an important step because it looked ahead to what happens when I am no longer guiding Safdie Architects—placing significant responsibility in the hands of five senior partners. I discounted the equity shares as much as possible, with a view to relieving financial pressure in the future. Nothing dampens excellence and creativity like the need to take on projects whether you want to or not. Among architectural firms, there is no single path to continuity. Sometimes, when a principal dies, the firm is successfully taken over by a partner—as happened long ago, after the early death of Eero Saarinen, when Kevin Roche and John Dinkeloo took the reins. Sometimes a firm's work has turned into such a "brand" that the firm becomes attractive to actual investors. Some architectural enterprises, such as SOM and HOK, are all-purpose corporate sauropods, like giant law firms, and can perpetuate themselves indefinitely. Some small firms are simply swallowed by bigger firms or fade away.

The Emperor's Landing, in Chongqing, China, as it looked in pre-Communist days (top) and then again before the current project.

A second consequence of Marina Bay Sands was that the project attracted the interest of CapitaLand, the biggest developer in Singapore and one that operates not only there but all over China and many other countries in the region. It was headed by a legendary CEO, Liew Mun Leong. One day, in 2010, I got a phone call from Liew's assistant: "Mr. Liew and his team would like you to guide a visit to Marina Bay Sands." They had good reason to be curious, as CapitaLand had been one of the prime competitors for the project, and they were not happy when they lost it. Liew, then sixty-four, was one of the stars of the founding generation of Singaporeans, whose path parallels that of the state. His optimism and his determined approach to every undertaking brought results. He enjoyed the confidence of both the government and the business community. Liew had come from a poor family. Early on he worked as an engineer in the public sector, building runways for the then-new Changi Airport. With close ties to the political leadership, he took on CapitaLand, a real estate company that is partly owned by Temasek, the sovereign wealth fund of Singapore, and partly a listed private company. Under Liew's leadership, CapitaLand became the mammoth developer that it is today, building mixed-use "Raffles Cities" all over China. (The name Raffles derives from that of the iconic Singapore hotel, dating back to 1887, and famous for the Long Bar. Sir Thomas Stamford Raffles was the British colonial governor who founded modern Singapore early in the nineteenth century.)

Liew was impressed not only with the audacity of the architecture at Marina Bay Sands but also with how well everything was functioning. The fact that Marina Bay Sands was making money also caught his attention. Liew immediately brought us in on two new projects: one, a middle-income residential complex on the outskirts of Singapore; the other, a competition for a 12-million-square-foot development in Chongqing, one of the most populous city in China.

The residential project, Sky Habitat, in Singapore, was soon designed and built, and opened in 2015. It is today fully occupied. It consists of a pair of stepped towers, thirty-eight stories tall, linked by sky bridges on three levels. It offers terraces and balconies for all units and generous public areas. The three sky bridges provide communal spaces, playgrounds, gardens, and swimming pools for five hundred middle-income families.

The project in Chongqing at last offered us a chance to break into a country that had long proved impenetrable, with its opaque governance and mysterious decision-making. Liew believed that the Chongqing project, located on historic Chaotianmen Square, was potentially the most important project in China. Soon he and I were on location, strolling around, trying to get the lay of the land. It was a glorious site, the tip of a peninsula formed by two great rivers, the

An early sketch from 2011, inspired by local sailing ships, of how that same space in Chongqing might be redeveloped.

Yangtze and the Jialing. One can see the different colors of the two rivers, brown and blue, as they merge.

In my enthusiasm, I had to remind myself that this was a design competition, and that we had not fared well in China with competitions. I had been back to China in 1999, many years after my first visit, in 1973, and had eventually begun participating in design competitions for major cultural projects—a performing-arts center, a science museum, an educational facility, an art museum. I remain proud of our design work—we keep the models for some of these projects on view under Plexiglas in our office. But we had entered these competitions with a measure of naivete, not realizing that China is all about personal relations and lobbying and holding hands. The results were discouraging. I wondered whether this new project would be any different.

In character, the city of Chongqing lacks the sophistication and worldly renown of Shanghai or Beijing. It is far inland—some fourteen hundred miles from the sea—and is profoundly Chinese. The city's people are much less exposed to the West. As I walked the streets, it seemed as if half the population had just arrived from the countryside—which it had. It is a city of gleaming new buildings and also of street vendors and traditional markets. The sidewalks are clogged with merchandise. Elderly laborers pull or push their heavy carts. Street vendors everywhere sell dumplings and noodles. At the same time, Chongqing is a center for many high-tech industries. Hewlett Packard is there, along with Toshiba, Sony, and Foxconn. Ride-hailing services are a click away, but walking is always the fastest way to get around.

The site being developed in Chongqing was the equivalent of lower Manhattan: the very tip of the city, known as the Emperor's Landing, because long ago an emperor had made his way up the Yangtze from the coast, disembarked at this place, and founded the city. In effect, the Emperor's Landing was the gateway to Chongqing, and it was already the site of a prominent plaza that was to remain in place. But the government wanted to develop the area in a major way and to make the site at the confluence of the rivers a showcase. It was a megaproject of great significance.

In China, a design must always be accompanied by a narrative. The Chinese prefer to relate to a design with a backstory that reinforces their thinking. This need goes far beyond the demands of feng shui, and it seems to apply to everything. When you eat something in China, you are given information about how a particular ingredient favorably or unfavorably affects various parts of your body—your eyes or your heart or your bladder.

In this case, thinking of a narrative actually helped inspire a design. The government had said it wanted a single primary tower at the tip of the peninsula, to define the site like a stake in the ground, with smaller, secondary buildings trailing behind or to the sides. I felt that that was the wrong approach. Instead, because the site served as kind of entryway, leading to the rivers from the city and to the city from the rivers, I envisioned paired towers that together formed a gateway. But this in itself did not hold the project together. The desired density required eight towers, to provide the needed 12 million square feet. How to give such an assembly any kind of larger meaning—something other than building towers more or less side by side—became our goal.

The word "landing" in Emperor's Landing provided a clue. For countless generations, merchant ships with their stacked rectangular sails had plied these rivers, landing at the emporium where the rivers met. Maybe the eight towers could form something like the prow of a sailing ship—the two gateway towers in front, the others arrayed behind. In the design, we gave them directionality, bending them dynamically as if they were billowing slightly. The two gateway towers would be the tallest, each with seventy floors, and the others would splay behind and to either side. Besides curving the towers, we linked four of them at the forty-fifth floor with a horizontal structure, a glazed, tubular sky park we called "the conservatory." It spanned the four "masts" like a yardarm.

The design and the narrative clicked—the emperor, the rivers, the gateway, the sails of a ship. Importantly, it clicked with the most influential persons in Chongqing at that moment: not just the mayor but also Bo Xilai, the city's Communist Party chief and a member of China's politburo. (Bo would soon run into trouble—his wife implicated in the murder of a British businessman—and is now serving a life sentence in Qincheng Prison, in Beijing.) Narrative aside, Chongqing represented a quintessential megaproject. Its 12 million square feet included two million square feet of retail space and required integrating three distinct modes of transportation: buses, subways, and ships.

The complex presented a particular challenge in terms of its relation to the surrounding city because it separated the downtown from the historic Chaotianmen Square at the tip. It was also essential that the shopping complex in the podium connect seamlessly to the city, avoiding the "inward-looking mall" syndrome. We therefore extended the surrounding streets into the complex, but pedestrianized them so that they became daylit gallerias flowing from the city to Chaotianmen Square. As one might imagine in a project of this scale, the budgeting and finances—all in the hands of the developer— represented a challenge in themselves and were constantly moving targets. Developing Chongqing required a total investment of nearly $4 billion—a figure that is misleading. Had Chongqing been built in Manhattan, the investment would have approached $10 billion.

One great advantage of working in China is the unlimited labor and the Chinese ability to manage and execute the work. The project in Chongqing required more than five thousand laborers, which in turn required building the equivalent of a small town's worth of dormitories. But this is nothing in China, which seems to be able to open a new university, on average, every couple of months. The abundant labor, coupled with an ambitious and adventurous spirit among the decision-makers, explains why so many leading architects worldwide are sought after in China. The project in Chongqing needed four general contractors—it was too big for just one. Michal, who often accompanies me, is entranced as a photographer by two elements at the core of construction activity wherever we go: the heroic, Meccano-set forms of the structure taking shape, and the lives of the workers themselves, most of them from rural areas, drawn temporarily to a city for the opportunities it affords.

In some ways, the most aggravating aspect of the project in Chongqing was not the work itself but the lawsuit brought by Sheldon Adelson in 2012. The conservatory that linked four of the Chongqing towers was not unlike the SkyPark at Marina Bay Sands in terms of its function, but it was a glazed and enclosed space rather than an open park; and it was a tube, not a landscaped platform. And it did not rest flat on top of the towers. To see it lifted in place in sections felt like something out of science fiction. The design was published in professional magazines, and it caught Adelson's attention. He called me: "You're using *my* design for another project," he told me. I said: "What do you mean? Obviously there are common elements—it's the same architect—but it's a totally different building." He said: "It's too much like mine—I don't like it."

I remembered that we had made sure to keep the copyright for the Marina Bay Sands design when we negotiated the contract with Las Vegas Sands. Like all developers, they asked us to give them ownership of the design (with all the implications), but unlike many architects, we always insist that we never give up copyright. In turn, they asked to add a clause stating that we would not use the identical Marina Bay Sands design in another location: "Safdie USA agrees not to utilize or permit to be utilized the overall design for the entire Project (except as part of this Project or in attribution) for any other project or in any way, without the prior written consent of the Owner." To us, the Chongqing and Marina Bay Sands designs were vastly different; this was self-evident. We were not in any way using "the overall design for the entire Project." But the meaning of the inserted clause became the subject of a legal battle that went on for fourteen months and cost us $1.5 million in legal fees to defend ourselves. Adelson sued us for $50 million. He also put a hold on the $2.5 million he still owed us and tried to bring an injunction to halt construction in Chongqing. Fortunately, Liew Mun Leong stood by us, and in the end, an arbitrator was appointed by the courts. The argument I made was this:

> The work of an architectural practice is, by nature, evolutionary and will inevitably contain recurring themes, elements or forms that may reappear, but also transform and develop over time across many projects. Each work builds on the works developed before and seeks to continue the exploration, albeit in a different setting, with a different program and for a different client. It is unreasonable to think that future works will not bear some semblance to past projects.

The arbitrator agreed. Adelson was instructed to cover our legal bills and pay us the fees that he owed, which he did.

Years went by without any contact with Sheldon or his company. Then I began to receive feelers about a reconciliation. First, the ubiquitous Rabbi Shmuley Boteach, who was close to Adelson, gave me a call. "Two prominent members of the Jewish community should not be fighting in public," he said. I told him that I held no grudges. I was open to receiving an apology. The next emissary was George Tanasijevich, the Singapore-based president of Marina Bay Sands, with whom I remained on friendly terms throughout the lawsuit. "Moshe, we need you; it's time for reconciliation," he said. Two new projects were in the works: an integrated resort in Bangkok and a major addition at Marina Bay Sands, which would include a new tower and an arena. The thought of another architect designing an addition to Marina Bay Sands was not appealing, but Michal would have none of it: "You are not going to work with them again. Have you forgotten the sleepless nights? The hours conferring with lawyers in Singapore? His four legal firms against our one? The financial hardship? All the money he held back? The bullying and intimidation?"

Eventually she mellowed. I drafted an iron-clad memorandum of understanding to govern future relationships, stuffed with every protective clause that I and our lawyers could think of. A few weeks later, when I was again in Las Vegas, I met with Sheldon. He said, "Let bygones be bygones"—not exactly an apology. His wife, Miriam, soon joined us. She whispered in my ear: "So, did he apologize?" I repeated what he had said, noting that it was not exactly an apology. To which she responded, "For Sheldon to ask you, having lost the suit, to come back—that, Moshe, is the greatest apology." We are currently at work on the Marina Bay Sands addition.

* * *

In 2012, as Chongqing was rising on its prow, CapitaLand came to us with another idea. Liew, the chairman, was moving over to become the chairman of Singapore's Changi Airport. CapitaLand had decided to compete for an airport project announced by Changi. Three other developers, each with its architect, were short-listed. The winning developer would enter into a joint venture with Changi Airport. The essence of the project was to create a new center, in the middle of the airport among the three existing terminals, unifying the complex. It would be on the landside—that is, before people go through passport checks and security—and hence accessible to the entire population as well as to passengers. It would expand airport facilities and add a major retail complex. The project also had to provide a "major public attraction," the choice of which was to be determined by the competing developers.

We had already designed two airport terminals—at Pearson International, in Toronto, and Ben Gurion, in Israel—and as a result had relevant experience we could bring to bear. The Changi project promised to be something beyond the ordinary in both scale and conception—as indeed it has turned out to be.

The task before us was not just to design an immense complex that linked all the existing terminals but also to create a destination in its own right. It was all about "Singapore meets the world. The world meets Singapore." To explore the issue, CapitaLand brought in a number of experts from the United States. Ideas were put on the table. All of them were some form of entertainment in the tradition of Disney or Universal Studios theme parks. As we learned later, our competitors all went down this path. One of them proposed a Dinosaur Land.

With Liew Mun Leong, onetime CEO of CapitaLand and chairman of Changi Airport.

I had strong reservations about that approach. Themed experiences are usually a onetime thing. They begin to feel dated quickly. Why would a passenger, or for that matter a Singaporean, be attracted to such a place repeatedly? The city had similar offerings already. We needed to come up with a unique concept that appealed to all ages, that appealed to passengers, that was timeless, and that made one want to come back. The idea of a magical garden began to take hold—a garden like no other in the world. I remembered, too, what we had done with the rotunda at Ben Gurion, and realized that we could push the concept much further.

I was imagining a giant dome made of glass and steel, hovering over layers of retail and other required spaces. The largest single feature would be an immense tiered garden, lush with vegetation. The glass dome would be shaped as a toroid, like a donut, so that its surface dipped toward the center, funneling rain—of which Singapore gets a lot. As worked out in the design, the rain would come down as a 130-foot-high waterfall—flowing at 10,500 gallons a minute—and would be recycled and pumped to remain continuous during dry spells. I thought the idea of a garden was appropriate because airports are harsh places. People are tense. If they spend any significant time in an airport, it's because they're stuck. A garden would offer relief and regeneration. It could be complemented by many mini-attractions: climbing nets, mazes, and other surprises. Passengers would go to it again and again. So would people from the city who weren't even intending to fly.

A notebook sketch, from 2012, shows the emerging concept for Jewel, at Singapore's Changi Airport.

We experimented with the concept, fine-tuning the geometry and details—I have some of the models hanging on my wall at home—but the reality, now known as Jewel, is almost surreal. The dome of Jewel is 650 feet across—five times the diameter of the dome of Saint Peter's—and supported in a way that seems to make it weightless. When you're underneath, you have a sense that the force of gravity has disappeared—except that water is falling continually from the top and into the middle of the garden, so powerfully that it creates a mist. It's as if you're out in nature, and in fact you *are* in nature. All the terraces are made of lava stone, designed so the stone itself gradually recedes in visibility, overtaken by trees and other plantings. The flow of the waterfall can be adjusted, like a showerhead. At times the flow seems the most natural thing in the word, as if one has parted leaves in the jungle . . . and there it is! Or some mechanical teeth can be projected into the flow at the top, which has the effect of making the downward flow so smooth that it becomes like a flat-screen TV and can display static or moving images.

* * *

Making Jewel a reality required backroom operations of a very technical nature and of a magnitude that members of the public can scarcely imagine. And that is probably as it should be. The innovative work of the structural engineering firm Buro Happold was essential to the Jewel project. Still, I sometimes get asked how we move from a general concept for a building, maybe scrawled in a sketchbook, to intricate, accurate drawings of a building's every last detail, down to the level of individual screws. This is, quite literally, the nuts and bolts of architecture.

Think of it as a process of escalating scales, where each scale widens the magnification—or as a camera with a lens that you can telescope. The initial rendering stage, based on knowledge of the building's purpose and program, gives a general idea of the envelope, the volumes, the relationships. Beyond the concept stage, we sit with the engineers and discuss: wood or steel, welded or bolted. We talk about glazing, about mechanical systems. Slowly there is a determination about each specific question: Yes, we're going to have air-conditioning, and we'll put the units in the basement, and the ducts will come up this way, not that way. And we're going to have windows here and here and here. A better picture of the materiality emerges.

The next phase is called design development. Here we have to figure out and draw, in general terms, how everything is to be fabricated and constructed. If it's a wood-framed window, we show the frame, the moldings, the glass, the hinges. We begin to understand all the pieces and how they work. And then we assemble them into a building, in the form of plans and a model. This is where the interferences are revealed—when we see how one system sometimes gets in the way of another system. The duct is supposed to go here and turn there . . . until we realize that, for some reason, it can't. Meanwhile, costs and budgets are assessed and reassessed at every stage, often with the help of professional cost estimators—skilled consultants brought in from outside.

In the not-too-distant past, we drew plans, sections, and elevations by hand. These are the three ways of representing

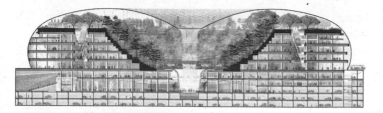

A more detailed cutaway rendering of Jewel Changi, 2014.

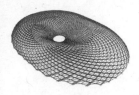

A computer-assisted-design image of Jewel Changi's asymmetrical toroid dome.

architecture. A *plan* looks at the organization of a building as if seen schematically from above—it shows the footprint. A *section* shows a building, or some part of a building, as if cut right through and opened up to be viewed. An *elevation* is what a building looks like on the exterior, from every conceivable angle.

In the modern era, plans, sections, and elevations can be created on a computer, and then the computer will convert it all into a three-dimensional image. One can travel through the 3D image and move about, look this way and that, and see it from outside and above and below—it is virtual reality in action. The software is now so sophisticated that, working independently on the computer, the structural engineer can put in the concrete and the steel, and the mechanical engineer can put in the pipes, and then all the elements come together—at which point that protest about a pipe going through a beam may still be heard. But it can be fixed on the screen, without redrafting everything by hand. And one almost always sees the problem coming in advance. In the old days, it often came on-site, during construction, when someone in a yellow hard hat tapped the architect on the shoulder.

This design-development stage is not the end—not even close. Once one has all the elements in place, defined to the last detail, then the world of the contractors and the fabricators takes over. The fabricators make specific products—a window system, for example—based on a specific technical design. The contractors and fabricators don't build from the architect's drawings; they cannot trust those drawings, because the drawings don't reflect their own fabrication methods. They look at the drawings and then adapt them to their products and hardware systems, producing what are called shop drawings. These are very finely detailed, and are submitted to the architect to check design intent and technical performance. We might look at the drawings and say, We don't like this because in a high wind it will leak; the wind will drive the water, and there's nowhere for it to drain out. Or we might not like the size of a window frame, feeling it's too heavy, and ask the contractor to make it more delicate.

In the case of Jewel, as the center of activity shifted to Singapore, the design team in Boston traveled to Changi to attend coordination meetings involving our Singapore staff and the client. Because of the culture of inclusiveness in Singapore, these were very large meetings—thirty or forty people in the room and a dozen active decision-making participants representing the client team (headed by Lee Seow Hiang, Changi's formidable CEO) and the design and consulting teams. I, together with Jaron Lubin, who headed the design team in the home office, and sometimes two or three others, as well as Charu Kokate, who heads our office in Singapore, attended each meeting.

When building something very complex, like the curved glass-and-steel roof on Jewel, the process jumps to yet another level. Jewel's dome holds some nine thousand panels of glass, each weighing upwards of 500 pounds. There is no way we could have designed Jewel's roof in the traditional way, with drawings on paper. Only computers could handle its complexity. In this new world, the fabricator embraces our 3D drawings and builds upon them. We can communicate from our computer drawing to their computer drawing, and vice versa. The computer can direct the machine that does the fabricating. But just checking the contractor's drawings is unimaginably time-consuming. For Jewel, this process of intensive review, which was done on a rolling basis as construction was underway, involved ten people in our Singapore office working full-time for four years.

For Jewel, we also built many full-size mock-ups. We built a small section of the dome, roughly thirty feet square, holding several of the glazed panels, to check the marriage of glass and frame, making adjustments and then checking again. We built a four-story-high mock-up, at full scale, of the four-level garden: the terraces, the irrigation systems, the stairs, the railings,

Mock-ups were essential for every aspect of Jewel. This full-scale mock-up tested glass for the dome.

the lighting, and every conceivable electronic system. We built mock-ups of the pedestrian tubes that would connect the garden to the terminals. All these mock-ups were so large that we had set aside an outdoor site just for them. We used a warehouse for mock-ups of floor finishes and smaller features. All in all, we built approximately thirty mock-ups of various sizes and complexities.

In parallel, our landscape team, Peter Walker and Partners, based in San Francisco, went into high gear to realize the concepts so elegantly rendered. For Jewel, which called for the planting of two thousand trees and one hundred thousand shrubs from all over the world—120 species in all—this meant countless international trips to select and tag trees and other plants and bring them to Singapore so they could undergo conditioning under tents and in greenhouses before taking their rightful places inside Jewel. Garden maintenance on this scale, indoors, calls for irrigation and fertilization systems, plus good drainage, and also misters and leaf washers to keep the trees fresh and healthy. Another team, Atelier Ten, an environmental consultant, dealt with atmosphere, climate, and sunlight. Plants need enough sunlight to thrive but not so much that they roast. Temperature and humidity need continual monitoring and adjustment. A balance must be maintained between comfort for plants and comfort for people.

Architects who forgo the painstaking supervision and review of shop drawings never get the building they think they're getting. We need to oversee every phase. In a major undertaking such as Jewel, some forty different consultant teams were involved. On my staff alone, at the home office and in Singapore, perhaps forty people participated in the planning and construction; another sixty-five people came to the project from CapitaLand and Changi Airport. The actual building of Jewel required fifteen thousand shop drawings, twenty-five hundred construction workers, and 26 million person-hours of labor. The design and construction

The central waterfall at Jewel Changi.

process played out over a period of six years and cost a total of $1.3 billion. Personal involvement with everything, and at all stages, is a lesson I learned first when working with Louis Kahn. The experience of every project since then has reinforced that lesson.

All this unseen work is in the service of something larger than any of the parts. Perhaps the word "magic" deserves another mention. It is a quality that Jewel certainly has. Some 260,000 people visited Jewel on the first day it opened informally, April 17, 2019. These were visitors from Singapore, not air travelers. Two days later, Jewel had 600,000 visitors. Jewel welcomed some 50 million people in its first six months. An art and cultural critic for the Singaporean website *Zula* wrote that she came to Jewel ready to hate it—another mall! more excess!—only to conclude that Jewel was "a game changer for a tiny city that everyone else around the world thought was a part of China just a year ago." She acknowledged that seeing the reality made her proud to be a Singaporean. (Actually, she added an emphatic expletive before "proud.") For a period of time, the pandemic cut air travel to (and through) Changi to 0.5 percent of its typical volume, and at one point the airport was momentarily closed. To experience Jewel when it was empty during this time must have been haunting.

When I show people pictures of Jewel, the reaction is often that they must be looking computer-generated imagery—renderings of something not yet built. There's a perception of unreality. But they are actual photographs of an actual place, undoctored, unenhanced. What fascinates me about Jewel is the way it hints that there's a way in which dense urban life can be full of nature—that polar opposites, the city and the garden, can be fused into a singular experience.

Study models for Habitat of the Future piled up in the model shop at Safdie Architects, Somerville, Massachusetts.

CHAPTER TEN

What If?

I remember watching, years ago, Leonard Bernstein's famous appearances in the Omnibus television series, in which he took viewers into the world of music—music of all kinds, from jazz to classical to Broadway musicals. And I remember his Charles Eliot Norton Lectures at Harvard, in which he delved into the subject at even more depth. I am a music lover, but these public commentaries provided tools and perceptions that immeasurably enhanced my knowledge and appreciation. I have often found myself hoping that a practitioner might one day produce something comparable for architecture. Ordinary people have an awareness of architecture—it is as ubiquitous as music—but not as much understanding as many would like. The one moment when people may come into personal contact with the architectural process, if they ever do, is in the design or rehabilitation of a house—and it is always eye-opening, even if they've spent their life as an architect.

In 1982, I bought from Harvard a house located on the Cambridge Common that had been the residence of Dr. Benjamin Waterhouse, the university's first professor of medicine. Waterhouse was a friend of John Adams—his Harvard roommate, no less—and had introduced the smallpox vaccine to the United States. The house dates back to 1753, and like all houses of that period, except those of the very wealthy, it had tiny rooms and low ceilings and small casement windows. It was built of timber. Because the house was a designated landmark and lay in a historic district, nothing could be done to change the building's envelope, though there was some flexibility with respect to the interior. Surprisingly, the owner was allowed to make additions and changes to the back, as long as they were not visible from the street.

We decided to restore the house pretty much as we found it. Michal and I agreed that the dining room would be lit only by candles, to recall the time when the house was built, but we added a circular ceiling mirror over the dining room table to reflect the candlelight and to compensate for the low ceiling. The room was painted a deep Japanese red. We opened up the kitchen to the garden in back, and then, to make up for all the low-ceilinged rooms, we added, almost as an attachment, a double-height living room that jutted dramatically into the garden. We planted some new trees to join the majestic ones that were already there and dug a small pond at the garden's end.

I was somewhat intimidated by the idea of adding to a 1753 house in a historic district, so the addition was subdued. The big windows and glass doors facing the garden were subdivided into smaller panes, in classic Federal style. The roof over the living room was shingled, and it sloped toward the garden. At the time, we thought it to be an exciting space, though I realized too late that the limited height of the glass windows and doors obscured a full appreciation of the trees hovering above.

Windows, like staircases, are a basic architectural feature—so fundamental to so many aspects of any building that they present a complex challenge. Do you want acres of glass or small apertures? Flat or curved? Hinged, pivoted, hopper, upward or downward, casement, double-hung, sliding, or double sliding? With or without screens? Wood, steel, aluminum, or a combination? Clear glass, reflective, tinted, translucent? How energy-efficient do you want or need the glass to be? And how easy will the windows be to clean—in a home, in a skyscraper? And how do you keep birds from flying into them? This last is a new, specialized discipline in its own right; some new kinds of glass are printed with ceramic dot patterns that people don't notice but are enough to warn off birds. Experiments go on.

The expanded living room overlooking the garden became a stage set, and was the scene for a moment I will never forget. Yo-Yo Ma and Michal celebrate their birthdays three days apart, and we have from time to time marked the occasions together. In 2006, with Michal turning fifty-five and Yo-Yo turning fifty-one, we decided to have a quiet dinner together at our home—ourselves and perhaps a few others. Two days before the event, Yo-Yo called to inquire about our piano. It had been in the living room but we had moved it to the second floor. Could you move it to the living room again—and have it tuned? Yo-Yo and his wife, Jill, and Michal and I had been attending Daniel Barenboim's Norton Lectures at Harvard that year, and Yo-Yo wanted to invite Daniel and his wife, Elena, to join us for the birthday dinner. Yo-Yo and Daniel had not played together for more than a decade. Might they give it a go? Of course, we said. After dinner, we retired to the living room. Yo-Yo suggested that they play Chopin's

Sonata in G Minor for Cello and Piano. He had brought the
music, because Daniel had not played the sonata since performing it with his late wife, Jacqueline du Pré. They sat side
by side, playing flawlessly without rehearsal. It was one of
the great performances—and for an audience of half a dozen.

In 2012, after living in the house for thirty years, Michal
and I decided to rebuild the addition. This turned out to be
an extraordinary lesson in the power of architecture—not
in the sense of dramatic form-making and grand geometries, but rather in terms of subtleties: how little details and
improvements could change the quality of life in a structure.
I was, at this point, less intimidated by the historical-review
board and perhaps emboldened by my experience of Jerusalem, where increasingly I had explored greater departures
from the local vernacular—trying to bring contemporary
design to an ancient city in a way that enhanced both. On
Waterhouse Street, we decided to rebuild the north-facing
living room and the attic bedroom essentially as glass-
and-steel greenhouses. In the living room, the traditional

View from our
living room to
the garden in
our eighteenth-
century home
in Cambridge,
Massachusetts,
after the first
intervention in
1982 (top) and the
second in 2015
(below).

The view from outside: our home before and after the renovation.

wood-framed glass doors gave way to large, plate-glass aluminum doors, which could be folded, accordion-style, to open up to the garden. Immediately above the doors, the glass continued, greenhouse-style, in a curved form. The swaying tops of the mature trees were now visible from inside the house—indeed, they sometimes seemed to have come indoors. We also added a south-facing strip skylight that let the midday sun into the room. The combination of northern light and garden plus warm southern sun made the space sublime.

And the garden can be seen from the kitchen, which opens onto the living room but is a few steps higher. After thirty years of our preparing meals for friends and family—food is a central preoccupation for both Michal and me—the renovation gave us an opportunity to rethink our kitchen design.

We also decided that the living room should have a fireplace, something we had missed in the space over the years. The east wall was reconstructed with heavy timber, which we obtained from a barn in Vermont. It was the same type of old pine used in the wide planks of the original house. The timber wall, with coves for sculpture, framed a large, black-granite fireplace. The combination of generous glazing, a view up into the trees, and a fire blazing, on a winter evening with snow falling outside, made me realize the extent to which this particular space had been transformed, though the dimensions of the space had not been altered. The light was superior, the contact with nature was intimate, and the presence of a fireplace anchored the room and gave it both gravitas and a quiet, flickering dynamism.

I was surprised but also edified by the experience. I had been building for decades, but this experiment, this little laboratory, which we knew so well, as we lived in it, demonstrated the truth of Mies van der Rohe's famous dictum: "God is in the details." The attention to materials, to light and transparency, and to harnessing appropriate contemporary methods made it possible to transform a Federal house with a pseudo-Federal addition into a home offering a rich, contemporary experience.

Even from the exterior, as one stood in the garden looking back on the house, the contrast of greenhouse architecture—large panes of curved and flat glass—in the overall setting of historic Cambridge seemed refreshing. I had not designed the new addition as an object to be seen from the garden. I don't think I ever studied an elevation as I worked on the house. The design was conceived completely from within, as I tried to imagine the experience of what living in such a space could actually be like. This, perhaps, is another demonstration of a fundamental lesson for architecture: Design is not only about imagining an exterior form. It is even more centrally about the internal spatial experience that gives rise to form.

* * *

Compared with other projects—for governments, for cultural institutions, for private clients—designing one's own home is an act of personal expression in which an architect has enormous freedom. As always, there are resource constraints. The client can at times be hard to deal with, even when it's oneself. But the possibilities lie substantially in your hands as an architect. Sometimes, I start to wonder about what I might do if I had similar control over the larger world—if I could decree improvements of my own devising.

Recent events have highlighted the need to set aside business as usual. The Covid-19 pandemic and lockdown devastated most major cities and threw into even starker relief the enormous divergence of income and opportunity in our society and worldwide. It also revealed the vulnerabilities of basic urban infrastructure. Meanwhile, families by the tens of millions, many with young children, spent months confined to their residences. A friend—an accomplished city planner—called me from her East Side apartment in Manhattan. Though she lives in a so-called luxury building, her apartment has neither a balcony nor windows that can open. Her message to me was simple: all apartments must henceforth, by law, be naturally ventilated, and all must possess a balcony. Perhaps this experience of pandemic has taught us to appreciate basic standards for the amenities that impact our lives.

The pandemic has raised other questions. What proportion of a city should be open space, as opposed to buildings? When does high density become too high? Two decades ago, we typically accepted as the "standard maximum" a floor-area ratio of twelve in midtown Manhattan—that is the ratio that prevailed at the time. The floor-area ratio, or FAR, is obtained by dividing a building's total square footage by its lot size. It is a key tool of zoning, where zoning exists, because it can be used to regulate desired density in particular locales; it varies within any city and from place to place. (Buying "air rights"—in essence, when a developer pays a neighbor to stay low so that the developer can go higher than might otherwise have been allowed—is one way to get around any restrictions based on floor-area ratio.) The floor-area ratio in midtown Manhattan reflects the fact that although buildings are packed closely together, smaller buildings exist alongside tall towers; if every lot in midtown were built out to its full potential, the city would become unlivable. Then came the wave of needlelike, super-tall apartment towers in midtown Manhattan. Real-estate considerations are increasing urban density, unchecked.

Will we remember the lessons of the pandemic? Public memory can be all too brief. In 1973 there was an oil crisis; people everywhere waited in long lines just to get gas. Behavioral transformation was instantaneous. People thought twice about commuting; there was a moment of mass shuffling and internal migration within metropolitan areas. Compact cars started flying out of showrooms. And then we forgot all about it, and giant SUVs and trucks took over the market. People went back to commuting by car. So it may be with the pandemic. Or, perhaps, we will, in places, seek a better way.

Which brings me back to the question of what I would do if I could decree improvements on a large scale. As I think about the answer, I want to imagine freely—to think of projects and ideas that I hope and wish I might be able to undertake and help realize; or, alternatively, that will be embraced by coming generations of architects and city planners. These are not the remote fantasies of a distant future—a *Blade Runner* city of floating vehicles; a performing-arts center on Mars— but efforts we could start tomorrow, if only we chose.

* * *

A City without Private Cars

Every major change in the structure and form of cities was ushered in by a revolution in transportation. With the architect Wendy Kohn, I wrote a book in 1997 called *The City after the Automobile*. The book began by noting that cities everywhere had long since reached the limit when it came to private cars; they were being choked to death. New measures designed to solve the problem of the car (more highways, bridges, and tunnels; more parking) ended up making the problem worse; meanwhile, mass transit in many American cities was a catastrophe, compared to prevailing world standards, in part because of its limited appeal—the spoiled American driver insisted on having the freedom of a car. An Oscar-nominated short animated film called *What on Earth!*, released in the 1960s and now regarded as a minor classic, took the form of a Martian documentary about our planet, viewing it from spaceships. Metallic organisms traveling on wheels were perceived to be the dominant species. Each of these organisms, though, was afflicted by at least one two-legged parasite.

The City after the Automobile was an exercise in both realism and imagination. The book was centered on the idea that,

A rendering from 2009 of how a public car-dispensing system might work in an existing urban zone—Boston's Back Bay.

although we would not give up our desire for automobiles, with all the mobility and flexibility they provide, we might be willing to opt for a system of vehicles on demand, preserving personal mobility without the burden of personal ownership, parking, and storage. A regime of cars on demand has other profound implications. A system where vehicles are readily available everywhere, and disposable at will, would transform the interface between various modes of transportation, which today causes a major bottleneck when it comes to flow and mobility. The worldwide introduction of on-demand bicycles for pickup and disposal at will was both the inspiration and the model for a somewhat parallel regime involving cars.

Since writing *The City after the Automobile*, many of our predictions have come to pass. First came the car-sharing company Zipcar, and today on-demand vehicles of various kinds, and from various companies, are commonplace in cities everywhere. As so often occurs, innovation began in Europe, where planning has a long history, and the idea then spread elsewhere. But inroads have been slow. Owning a car has been an entrenched habit (and a status symbol) in the developed world for more than a century—and in the United States in particular. I have not been immune to it. Owing primarily to romance with the automobile, Americans spend more money on transportation than on food. Business investments in urban land and infrastructure are premised on a regime of private automobiles. Typically, about a third of an American city's land area is devoted to parking whether on the street, in open lots, or in garages. In Los Angeles, there are more parking spaces than there are households. And yet most cars are sitting idle 95 percent of the time.

Many place their hopes for a new vehicular regime on the advent of the driverless automobile. We already know that the technology exists, and some enthusiasts say we'll be well down the driverless path in just a few years. I am not an expert on artificial intelligence, or on the control systems that would be required to make driverless cars safe and universal, but intuition tells me that a driverless world is not years but more like decades away. Decades is a long time in the life of cities. Forty years ago, Chinese cities had no high-rise buildings at all; today, Chinese cities boast the greatest concentration of high-rise buildings the world has ever seen (and, not coincidentally, the greatest level of resulting congestion).

What I believe is needed to break the resistance to change is a large-scale prototype: one place, one city, where the option of privately owned vehicles is eliminated and the concept of on-demand vehicles can be tested and proved. To be sure, there would have to be delivery and emergency vehicles; Uber and similar services; vehicles for the handicapped and the infirm. But the best way to test and accelerate the idea of a city without private cars is to create an urban prototype. That's the way to figure out the logistics, the economics, and the control technology; that's the way to design an urban fabric in which airports, railway stations, and large buildings have all been designed to accommodate the storage and dispatch of on-demand vehicles.

Consider a typical public garage in a system, like today's, where individuals park their own cars. When all the ramps and the need for interior traffic patterns as well as the need for room to maneuver in and out of spaces are taken into account, the average car, which takes up a little over six hundred cubic feet, requires twenty-eight hundred cubic feet of garage space, a ratio of more than one to four. An automated storage-and-dispatch system that dispensed rented cars (at random) at the press of a button—like a Coke machine—would point the ratio in the direction of one to one. By itself, such a new storage-and-dispatch system gives you a sense of how radically different life could be.

For designers, creating this urban prototype would mean going back to first principles. How do you enter a building that does not have a massive garage but just a small, built-in silo in which vehicles are taken away and stored? What will it mean to design an airport, or a railway station, or any other place in the city with a need for seamless transfer from one type of conveyance to another? How do concentrated places of shopping, whether malls or marketplaces, become transformed when access no longer requires massive, tiered parking structures or vast tracts of asphalt parking lots, with thousands of people in a state of maneuvering, nerves frayed, looking for a place to put their cars? And there are larger questions. For instance, would a cars-on-demand system encourage greater dispersal of the city; or, inversely, with concentrations of on-demand terminals around rapid-transit stations, will it lead to greater densities?

The opportunity to realize such a prototypical experiment—one in which a single major variable is being altered—is in those places where entire new cities or new sectors of cities are currently being built. China would be the obvious choice. China is leading in the manufacture of electric vehicles. It is still in the process of building brand-new cities—about four hundred of them are under construction or on the drawing board. It has the will and the technological sophistication to lead in the area of urban transportation. Like any project of this transformative nature, it would require the collaboration of the public and private sectors.

This prototype city would spin off experiments in urban organization. I believe that an ideal city should provide at least one-third of its area as open park space. Add up every parcel of green space in Manhattan—every "vest pocket" park no bigger than a building lot; every shoreline bike path; every planted triangle—and one finds that only about a quarter of the island falls into that category. Now imagine a Manhattan where the proportion of green space that one finds in the Central Park latitudes was extended to the rest of the island. Parkland would take up well over a third of Manhattan. This is the proportion of park to city that would allow high- and medium-density neighborhoods, along with vibrant commerce, to be within walking distance of major parks and nature everywhere.

A city without private cars can be efficiently organized into a handful of connected centers, perhaps six or seven of them, each the scale of a downtown Boston or Fort Worth. Liberated from traffic and parking, these pedestrian downtowns would generate a rich and varied public and commercial life—a revival of the pre-automobile era. The various centers would evolve differing specializations and characteristics. They would be organized like a necklace, linked by rapid transit. Since the transition from on-demand vehicles to mass transit would be seamless, it would be easy to move from one center to another. Radiating from each center would be medium- to low-density housing, offering a range of densities and typologies—and therefore lifestyle choices—in the city. These would be served by the on-demand vehicles (as well as by on-demand transit), linking back to the center and from there to the rest of the larger city. The park system, too, would form a continuous necklace, just as the urban centers linked by transit do. Think of this as the "braided city," where all the systems—open parks, dense clusters, dispersed development, transportation arteries, even agriculture, including vertical farms—weave among one another continuously.

The braided city is a way of integrating nature and architecture to the point where the distinction between city and country, or nature and not nature, begins to disappear.

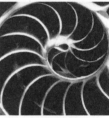

The chambered nautilus shell—an example of gnomonic growth.

* * *

Designing Nature's Way

H alf a century ago, in my book *Beyond Habitat*, I looked forward to a day when we might have something that I called a "magic housing machine." Not really imagining that we would actually have such a thing anytime soon, I wrote:

> *Heretofore we have thought of building in terms of the technology of today—the stamping machine, repetition. But the technology of building will become all-capable, like a computer punch-card with millions of possibilities extended in four dimensions or fluids capable of limitless forming. Ultimately I would like to design a magic housing machine to do just that.*

Trees transform
seasonally. Why
can't buildings
similarly follow the
cycles of nature?

*Conceive of a huge pipe, behind which there is a reservoir
of magic plastic. A range of air pressure nozzles around the
opening control this material as it is forced through the edges of
the pipe. By varying the air pressure at each nozzle one could
theoretically extrude any conceivable shape, complex free forms,
mathematically non-defined forms.*

When I wrote this, there were of course computers,
but the PC did not yet exist. Computer-aided drafting had
not yet been invented, much less 3D printers or the many
processes available for computer-aided manufacturing. My
magic machine seemed far-fetched, even to me. And yet,
by the turn of the millennium, we were well on the way to
having this very thing. There are investigations today not
only into expanding the range of traditional materials but
also into semi-organic "intelligent" materials, whose properties and behavior can change automatically, without human
intervention, under the influence of environmental factors:
heat, light, humidity, even magnetism. Neri Oxman at the
MIT Media Lab has pioneered the application of some of
these at small scale. But think of them at large scale—how
they might transform, for instance, the familiar glass tower.
Imagine if the properties of the sheath could adjust themselves incrementally in response to changing conditions
on the exterior.

My aspiration has always been to find a way in which
to transform the design process so that it has the kind of
organic fitness that is always present in nature's designs.
The typical glass office or residential tower building that
occurs in hundreds of cities around the globe—in scorching
Dubai, in freezing Montreal, in tropical Jakarta, in drizzly
Seattle—is badly in need of transformation. These towers are
basically the same wherever they are, no matter the season
or the climate.

By comparison, the design of a tree—a deciduous tree,
common to temperate climates that experience seasonal
changes—allows it to transform from its full leafy version to
a skeletal version, and then come back again. In hot weather
the leaves rotate to show their silvery reflective side. In harsh
sunlight, some cacti rotate to show a narrow edge to the rays.
I have long wondered why architecture's designs in response
to nature's forces seem so primitive in comparison to nature's

own designs. What if a tower could be designed to transform with the seasons, so as to absorb light and heat in the cold winter and to provide shade and reflectivity in the hot summer and even convert that summer sun to renewable energy? The objective of transparency can be maintained, but the building's envelope could be transformed to accommodate the seasons by using dynamic materials—putting nature's bag of tricks in the service of architecture. Thus, buildings designed for different places and environments could have the adaptive variability that different species possess in various climates. Steps have been taken in all these directions. At Crystal Bridges, for instance, we are experimenting with glass that can transform from clear to dark in the same way that a household dimmer can increase or decrease light along a continuum.

What might sound like fantasy—designing nature's way—could soon be a reality. The immediate applications can involve the automatic convertibility of spaces from (in essence) indoor to outdoor and not just on a seasonal basis but a daily basis, as dwellings adapt on their own from the sweltering tropical humidity of daytime to the breezy cool of nighttime. Imagine the concept of a convertible automobile extended to streets, playgrounds, public spaces, personal gardens, and roof decks. Building envelopes call out for dynamic adaptability in terms of transparency and permeability. We need—and there's no reason we can't have—a world in which architecture transforms itself as we watch it.

* * *

Habitat New York

For all the success of Habitat '67, most people have forgotten that the Habitat we were able to build was not the Habitat of the original proposal. The Habitat that was built for Expo 67 was only a small section of the original plan presented to the government in 1964. That plan proposed building a whole urban sector. It was not just residential but a mixed-use development in which all the traditional components of a city—residences, offices, schools, shops, restaurants, hotels—were three-dimensionally rearranged. Cities didn't have to be confined to towers erected on grids of streets.

Up to twenty-five stories high, the original Habitat—the version conceived, not the portion actually built—called for a network of streets in the air, supported by inclined A-frame towers with inclined elevators serving the various levels. A major continuous sheltered space was formed under these housing membranes, each uncompromisingly facing southeast and southwest for optimal orientation. The A-frames also supported a continuous roof promenade.

With time and resources limited, the Canadian government had decided on an arbitrary budget that allowed us to build only a small, twelve-story section of the whole. We would be able to demonstrate the "for everyone a garden" principle but not the larger concept of a reconfigured, three-dimensional city. I have often wondered what might have happened, in terms of urban design around the world, if time and resources had made it possible to realize the original Habitat concept. The interest we see today in rethinking the dense, high-rise city might have occurred decades earlier.

About fifteen years ago, I began to feel that much had changed and that it was time to reexamine some important ideas. At the time, we were building Marina Bay Sands and operating at a scale I had never attempted before. I realized that if I didn't take active steps to explore certain possibilities—about public housing, construction technology, and the creation of a public realm high above street level—the day-to-day pressure of the practice would ensure that I never would.

I was keen on the idea of reexamining Habitat and asking how we might do it differently today—with new technology, changed realities, profoundly greater population densities, and evolving ideas about urbanism. Urban visionaries and creative artists had been exploring these questions for a long time: witness the "stacked streets" ideas of the architect Harvey Wiley Corbett (1913) or the imagery in Fritz Lang's *Metropolis* (1927). I decided to set up a research effort in the office. We created a fellowship and assigned about $300,000 a year to bring two or three bright young architects to work with us. We looked at many topics: density, technology, geography, climate, and construction efficiency.

To my mind, the most interesting study turned out to be what we called Habitat of the Future. The idea had actually come from Donald Albrecht, the curator of an exhibition of our work called *Global Citizen* that had traveled around for several years. We developed a series of hypothetical designs where we explored what emerged when the variables were changed. Could one make the Habitat concept cheaper but still retain all the amenities—the level of privacy, the open gardens, the multiple orientations, the public character of street access? What would happen if it were even denser? We produced dozens of studies, picking the four that seemed most promising to develop further.

All my work is informed by the site and by the specifics of a location, be it the waterfront of Singapore or the ocher hills of Punjab. With respect to the next stage of Habitat, we decided to investigate whether Habitat on the Saint Lawrence River could be extended to an evolving waterfront site in New York City. How might one reproduce the mix and density of midtown Manhattan in a way that doesn't create dark canyons and that brings light and the lushness of plant life to every level?

For this imaginative exercise, the members of the fellowship program and I, along with several colleagues, started by analyzing a large swath of midtown south of Fifty-Ninth Street and establishing the overall density (a floor-area ratio of about twelve) as well as the proportionate mix of office, residential,

In 1913, the designer Harvey Wiley Corbett produced a vision of a city of the future.

Habitat of the
Future, an
imaginative
rendering from
2008. Projects
inspired by the
concept have
been built in
Qinhuangdao,
China, and
elsewhere.

commercial, institutional, and public space. With this data
in hand—and resolving to keep the density ratio and mix the
same—we embarked on reconfiguring a mixed-use, multilevel, three-dimensional city, seventy-five stories high. The
ground plane, familiar to us as "street level," would contain
the shops and traffic we're accustomed to, but reconfigured,
with weather-protected galleries and generous amounts of
landscape. Above, we stacked three clusters, each twenty-five
stories high. The first twenty-five floors would be devoted
primarily to office space, with a major public promenade or
"street in the air" connecting all the towers, one to another,
at the twenty-fifth-floor level—in effect creating an urban
canopy with its own circulation, services, day-care centers,
playgrounds, swimming pools, and shops. The next twenty-five
floors would be residential (including hotels)—at which point,
on floor fifty, a second horizontal streetscape in the air would
appear, likewise connecting the buildings but strictly devoted
to community space. Then would come the final twenty-five
stories of residences. The whole assembly would consist of
stacked sectional towers, the sections progressively stepped
back so that rather than forming walls, the buildings would
form giant urban windows twenty-five stories high and 150 feet
wide. In other words, the city would have porous massing. It

would be permeable, penetrated by sight lines—even from the interior, one would see the rivers, the bay, Central Park. The terracing would also create platforms of many different sizes for gardens, some of them public, some of them private, but with a profound overall effect: the hanging gardens of New York.

The design we arrived at spoke directly to what I believe must be the next stage in the design of high-rise towers—and flowed directly from one of the major breakthroughs of early twentieth-century modernism. Until then, domestic architecture, from the Renaissance palazzo to the Victorian mansion to the standard apartment building, was highly introverted. Interior space was clearly separated from the surrounding landscape, be it urban or pastoral. Architecture was shelter—meant to protect people from the elements (and from other people). But architects like Frank Lloyd Wright, Richard Neutra, and Erich Mendelsohn pioneered the idea of creating a seamless flow between inside and outside. Wright's so-called prairie houses are about a new way of living; the garden, the landscape, and the outdoors are integral parts of our living space. It was a revolution, but one that seemed to have no impact on the design of high-rise towers, which for the most part continued to be conceived as sealed extrusions. Sometimes balconies were added as a way to exit to the outdoors, but by and large, particularly with office buildings but also in most domestic towers, the old ways prevailed. This is the moment, reinforced by the pandemic experience of millions of city dwellers, for a new paradigm.

The first stage of Habitat Qinhuangdao, completed in 2017, two hundred miles east of Beijing.

With Habitat New York, we showed that a new kind of urbanism, incorporating communal gardens and personal outdoor spaces, could achieve formidable density and create many opportunities for meeting and interaction while avoiding the congestion and claustrophobia we associate with great density. It was like stacking one hill town on top of another but maintaining a clear hierarchy. Public spaces would tie the fabric together, and define the public realm.

One surprising outcome of the Habitat of the Future explorations was that we started to get commissions that enabled us to apply what we were learning—for example, Qinhuangdao Golden Dream Bay (also known as Habitat Qinhuangdao), in China, begun in 2010; and Sky Habitat, in Singapore, begun in 2009. Considerably greater in scale than the original Habitat '67, pushed skyward by a factor of three or four, and yet conducive to communal life high above the ground, these projects were offshoots, one way or another, of work that had been undertaken by the fellowship program. In these developments, we're already seeing the kind of rich community life that occurs on the bridge levels. I get letters from people living in them about the quality of their lives. Some may wonder whether the self-sufficiency of communities high above the ground may prove isolating— if "remoteness" will create a kind of social bubble. Work, schools, city attractions, and the vibrant activity of ordinary urban life will create all the necessary connections at every level, many of them taking forms we cannot now imagine.

A soaring, mixed-use, high-density Habitat—one that extends the public realm vertically and horizontally; that reconfigures the relationship among living, working, and shopping; and that integrates park spaces generously throughout the urban environment: this, for me, is the unfinished symphony, the project I yearn to see realized.

* * *

Taking the Future into Our Hands

This isn't a project in any conventional sense, though it may be more daunting than all the others. It is a wish to see the United States profoundly alter the way it thinks about the physical underpinnings of a functional society, and in particular the character of our cities—a wish to see America entertain ambitious dreams.

Let me start with a story—of my own firsthand experience with the failure of an ambitious dream, an experience that is symptomatic of a larger problem. One of the more fascinating projects I've ever been associated with—it involved perhaps the most visionary and costly physics experiment in history—was the design for the facilities and campus of the Superconducting Super Collider, or SSC, that was to be built beneath the dry flats of Waxahachie, Texas, south of Dallas. The aim of the $4 billion supercollider, lodged in a fifty-four-mile-long circular tunnel, was to unlock the secrets of the basic building blocks of nature and to resolve unanswered questions posed by particle physics. The campus was to be home to some two thousand scientists from all over the world. I had been introduced to the supercollider project in 1992 by a Harvard colleague and friend, the Canadian physicist Melissa Franklin, who in turn introduced me to Roy Schwitters, who headed the team of physicists in charge of the project. After being officially brought on board, I had visited the campus-like facilities at Fermilab, in Illinois; Brookhaven, on New York's Long Island; Stanford, in California; and CERN (the European Council for Nuclear Research) in Geneva, Switzerland, to prepare for what lay ahead.

Part of a tunnel excavated for the Superconducting Super Collider, in Waxahachie, Texas.

We made a model of the entire Waxahachie region. We marked the footprint of the tunnel and the points at which it would need to be serviced. One question was: What to do with all the rock and soil being excavated to create the tunnel? While rapidly calculating the amount of earth we would need to dispose of, what came to mind were the earthworks of the artist Robert Smithson, whose *Spiral Jetty*, in the water of the Great Salt Lake, in Utah, and many other landscape sculptures used fill to create intriguing forms. Here we would get it for free, and rather than truck it off-site we proposed piling up the residual earth into exciting geometric shapes

Room for
improvement:
floorplan of the
factory-like,
World War II–era
facility where
supercollider
scientists were
already working.

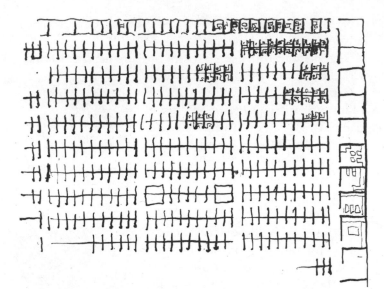

on the flat prairie, creating landmarks that would mark the
perimeter of the tunnel. Meanwhile, because the supercollider needed continous cooling, lakes and ponds would have
to be dug to hold ample amounts of fresh water. The lakes
and ponds offered a feature to design a campus around. We
put residences on one side of a long, thin lake and labs and
offices on the other, connecting them with a living bridge
laden with auditoriums, a library, restaurants, and other shared
facilities—a reimagining of Florence's Ponte Vecchio for the
age of particle physics.

None of this was built. In 1993, Washington abruptly
suspended the supercollider program, fearing escalating costs.
Some $2 billion had already been spent. About fifteen miles
of tunnels had been dug; they sit cavernous and empty today.
Leadership in particle physics passed to Europe, home of
CERN's Large Hadron Collider. As *Scientific American* com-
mented, on the twentieth anniversary of the government's
abdication, "The SSC was an epic project that ended in failure. The U.S. has yet to stride again [in] its own once prominent footsteps."

I have always wondered what explains the poor state of
planning, infrastructure, and maintenance of the public realm
in the United States. In this regard, the nation stands alone
among the developed and thriving economies: Japan, Canada, all of Europe. Our major airports—New York, Chicago,

Los Angeles, Miami, Houston—are a disgrace: inefficient, ugly, unclean. The airports in Singapore, Beijing, Shanghai, Munich, and Madrid lift the spirit; American airports do the opposite. Our train system is decrepit and pales alongside the bullet trains between Shanghai and Chengdu, Beijing and Qinhuangdao, and Hong Kong and Shenzhen and Guangzhou; not to mention something as ordinary and efficient as the railway systems of Western Europe. Even worse, we can't keep what we have, no matter how old or new, in a reasonable state of maintenance. Bridges are in disrepair. Highways are being sold off into private hands. Rest areas are strewn with garbage. The sense of civic pride has vanished. So has the capacity to dream big.

The environment in which we live is, in part, the incremental product of countless individual actions and interventions. One should not minimize the significance of these individual acts and their ability, collectively, to have an impact on our quality of life. As I walk through cities and small towns alike, my eye relentlessly and critically examines every building I pass. When I see banal apartment buildings or schools or post offices, I can't help but wonder, "What might this have been?" or "How could this be improved?" When we go to Venice, to Rome, to Dubrovnik, we love every moment. The cohesiveness, the interplay of water and buildings, the richness of detail, the overall beauty of the collective. We look at these places as if another species, not human beings, had built them—some different species that somehow knew how to combine ambition, ingenuity, taste, consistency, and discipline, and could stick with these qualities over time. Who were these creatures—and whatever happened to them? The cities we like the most—places like Amsterdam and San Francisco and London—owe something to that vanished species. The more contemporary our cities are, the messier they are— Houston, Dallas, the new cities of China.

Some of those who bewail this state of affairs believe the answer is to set the clock back. For them, nostalgia is not a form of wistfulness but an actual master plan. Prince Charles, a prominent backward-looker, once asked: "Can we not learn from the age of Wren, that unique moment in our architectural history when the vernacular gothic and the classical were fused in a vigorously attractive style? Do we still have to strive to be a stunted imitation of Manhattan?"

A model of the
scientific campus
for the planned
supercollider,
1993. Washington
cancelled
the program.

One can sympathize with his sentiments, and even share his architectural taste, but nostalgia ignores both the disturbing elements of the urban past (the sewage, the air pollution, the overcrowding) and the developments of the past century that have altered cities forever. These include demographic and economic factors (the massive influx of the world's population from the countryside to the city, which will not soon be reversed) and the impact of technology (the advent of the automobile and the ability to build towers of ever-increasing size).

Physical realities are not the only things that have changed. So have ideas and attitudes, which can be as strong as steel and are harder to bend. One of these ideas is the notion that the marketplace, by itself, is the only regulator we ever need. If something happens because market forces caused it to happen, then the outcome is by definition good and even natural. I think often of the words of Octavio Paz, speaking about another subject in a way that applies equally to society as a whole: "The market, blind and deaf, is not fond of literature . . . and it does not know how to choose. Its censorship is not ideological: it has no ideas. It knows all about prices but nothing about value." Around the country, while city-planning departments, which deal with zoning and urban design, have been weakened, the influence of developers has only grown. The uncontrolled consumption of land, the neglect of infrastructure, the privatization of the public realm—all are manifestations of blind faith in the marketplace.

Inextricably related to this ideology is the retreat from public responsibility for what we hold in common. Starting with the Reagan administration, and extending through every Republican and Democratic administration since then, the general trend in the United States, at the federal as well as local level, has been to push for smaller government and to discredit taxation and regulation as being unfair and unwise—inhibitors of economic growth. The Tea Party may have sputtered out as an overt political force, but not as a state of mind. Tax levels have been continuously cut. California, for decades, had tax limits that made the state's responsibility for infrastructure and education impossible to shoulder. Americans once were capable of great things. In the United States, the last period of major commitment to the public realm, and to infrastructure more generally, was the one that began under Franklin D. Roosevelt in the 1930s and ended under Dwight Eisenhower in the late 1950s. The New Deal brought a major public investment in urban improvements and national infrastructure: the building of bridges and dams and government buildings all across the country. The last significant burst of activity was the construction of the interstate highway system, beginning in the 1950s. This kind of investment made with the future in mind is absent from public thinking in America today.

Two things are needed if the United States is to regain its capacities and restore its ambitions. The first is to admit the problem and change our outlook. Addressing the malaise must begin, of course, with political change, with an electorate that will demand programs of expanded infrastructure and investment in the environment and its maintenance. Leadership is one element of what is necessary. How often have we seen new administrations declare a commitment to infrastructure, only to have the commitment turn into a punch line? (The ambitious plans of the Biden administration may prove to be an exception, though it seems unlikely that they will mark a revolutionary turning point.) But it also requires a change in public outlook.

I remember a very different attitude toward the future—and this was not so long ago. When I was a youth in Israel, a favorite pastime every spring was to hike in the countryside and pick flowers. We would carry home buckets filled with narcissus, poppies, iris, and cyclamen, and these would decorate our homes. By the 1950s, flower picking had depleted Israel's wildflowers, to the point of near extinction for some. To the rescue came the political and educational leadership. A major public-information campaign reformed public understanding and behavior. Picking a wildflower became something that simply was not done. The landscape was restored. This is a small example of the kind of behavioral reform that can be achieved when there is conviction.

So, a change in outlook is essential—a prerequisite. The second thing we need is a renewed faith in deliberate planning. Big plans—plans that incubate and brew somewhere in the strata of power—have an outsize influence on our lives. They can of course be destructive. They can also do enormous good. Planning is what produced our national parks, our great city parks, and the City Beautiful movement (which gave us the parks and beaches of New York, Chicago, San Francisco, and Los Angeles). And lack of planning can do as much harm as bad planning. Three decades ago, if one looked from Manhattan across the East River toward Brooklyn and Queens, the cityscape in those boroughs consisted mostly of low-rise buildings. Except for downtown Brooklyn, there was not a single tower. Look in that direction today, and scores of towers are visible. They all went up rapidly and without much planning. All this development could have been done differently: there could have been a concept; there could have been provisions made for infrastructure; there could have been a plan to give the riverfront back to the people; in short, there could have been thought and foresight. But generally, throughout America, each developer gets a piece of land to control and then is allowed to maximize what can be built on it. These developers operate outside any broad cultural agreement about what is desirable, even when any such agreement exists, which is rare.

I remember visiting Robert Moses in the mid-1970s, toward the end of his life. He had invited me to his office on Randall's Island, in New York—in the shadow, literally, of the Triborough Bridge, one of his creations—because he wanted to know more about Habitat. He gave me a copy of his book *Public Works: A Dangerous Trade*, an autobiographical miscellany, and inscribed it warmly. Perhaps fittingly for someone with pharaonic dreams, his office seemed too big for him. There was still a spark in his eye, but his manner

was that of a lonely man and, perhaps, a bitter one. Robert
Caro's *The Power Broker*, recently published at the time of my
visit, had badly damaged his reputation. Jane Jacobs's earlier
book, *The Death and Life of Great American Cities*, presented
a stirring vision of urban life that could not have been more
different from that of Moses—and my sympathies lay with
her, up to a point. By force of will and eloquence, Jacobs had
saved Greenwich Village from his bulldozers. But a collateral
by-product was to discredit the idea of Big Plans altogether.
Moses advanced some destructive ideas, but he also gave New
York needed parkways and wonderful beaches and parks.
We need ambitious dreams. We just have to make sure our
dreams strike the balance between the concerns and passions
of a Jane Jacobs and the vision of the Big Planners—the great
challenge of urban planning.

Regulation is a crucial tool. It is instructive to observe
how cities that have regulatory stipulations about daylight
and sunlight penetration have been able to keep a handle on
density, massing, and building typology. Many of the major
historic cities of Europe are examples; in fact, it is mainly in
Europe that one finds such regulation today. The building of
the fifty-nine-story Montparnasse Tower, in Paris, in 1973,
jolted the city into quickly banning any further construction
above seven stories within its limits. The old joke about
Montparnasse is that its observation deck offers the best
way to see Paris, because when looking at the city from the
deck, you can't see the tower. When density, massing, and
typology are compromised—as has occurred in most of the
world—the city is transformed, and for the worse.

Beyond preventive regulation—aimed at protection—we
also need affirmative regulation that makes good things
happen. Regulation can tell us what to do. Perhaps certain
streets should be arcaded, and individual builders will have
to provide for that, however else their designs might vary.
Perhaps all apartment units must have a balcony or other
outdoor space, however different one apartment building
may be from the next. Specific ideas will vary. The point
is, regulation can address qualitative issues, creating a
framework around which individual building efforts can
coalesce into a whole.

* * *

Architecture in Zero Gravity

Every architect, and for that matter, every engineer, learns that the process of design and construction is a singular and continuous struggle with the forces of gravity. Most of us take it for granted; it is a force we are born into and intuitively learn to live with. The bumper sticker "Gravity: It's the Law!" sums up the idea. When you design a building, your pen stretching a roof across a wide space, you intuit with experience, "Oh, this is too far." A column is necessary, and then a beam, and then another column.

I've often fantasized: What would it be like to design without having to contend with gravity? What kind of architecture would emerge? Some of our greatest inventions in architecture are those responding to the force of gravity. The arch spans masonry walls and can be extended into arcades and aqueducts. Vaults and domes span across houses, across mosques and churches, across huge stadiums. Flying buttresses support the soaring structures of Gothic cathedrals; great cables allow suspension bridges to cross rivers and valleys. Each of these building blocks of architecture has evolved to cope with gravity.

The idea of an architecture free of gravity first occurred to me some years ago, when an inquiry from the National Aeronautics and Space Administration invited participation in the design of a space station. In the end, nothing came of the venture, but I kept wondering what it would be like to design a house "out there," where gravity does not exist. The great designer Raymond Loewy, I later learned, had explored the question, creating interiors for Skylab, the first U.S. space station. Starting to sketch, I was overwhelmed by the implications. Every surface of a cube-like space could be imagined as the locus for a variety of activities. An eating station could be affixed on one wall, a sleeping room on the opposite wall, and a lounge area on a third. The distinctions among wall, floor, and ceiling would be eliminated. We're so used to our ambient reality, governed by the familiar laws of physics, that it's not as easy to imagine a space where all surfaces are available for any activity. And there's no reason to restrict oneself to a cube.

Architecture in space would necessarily have constraints. Pressurized environments would be needed in which you could comfortably remove your space suit. The space would need to be filled with breathable oxygen. The form that would most efficiently contain a pressurized capsule is a sphere. Gravity would not be present, but the necessity to create and sustain an atmosphere would still introduce discipline. Such a sphere could be studded on its perimeter with several projecting semispheres, creating what I think of as "coves"— places for diverse activities and diverse moods. Certain volumes and spaces could be stretched into the center, tied to the perimeter, for communal activity.

A house in space would need to deal with sunlight—not the atmosphere-filtered, relatively gentle sunlight of Earth but the harsh, direct sunlight of space. Shielding the home from the severe heat and glare of daytime would be essential, as would insulation from the extreme cold as the sphere moved into Earth's shadow, with the sun on the other side. Cooking would pose new challenges. Something as simple as boiling water looks very different when there is no gravity.

One thing we can say with certainty is this: architecture for a gravity-free environment is coming—someday. The irony is that this new freedom will bring its own limitations, often limitations we cannot yet imagine. Human beings have had millennia to figure out what works—for dwellings, workplaces, and entire cities, and to accommodate a vast array of behaviors—in a world with gravity. In space we would start almost from scratch. Once again, the architect would confront constraints—the nonnegotiable demands of a new reality—and once again those very constraints would present opportunities and inspire creativity.

Interior of the synagogue component of the proposed Abrahamic Family House, in Abu Dhabi.

CHAPTER ELEVEN

Faith and Peace

For as long as I have lived, the Arab-Israeli conflict has been a center-piece of my experience. It started long before my birth, in 1938, and it continues to rage. It could not have been sustained were it not for substantive clashes of interest. It is a conflict with deep roots, one that shadows the lives of everyone involved.

As an Israeli Jew, I might be expected to advocate for "my" side of the conflict. But I have always felt myself to be more of a mediator, a facilitator, and have tried to behave as such. I grew up in a socialist, Zionist environment, and continue to feel a powerful identification with Israel, so I often wonder how this more detached personal attitude was formed. No doubt, there are many Jews and many Israelis who yearn to resolve the conflict. My own perspective, however, originates in the belief that a peaceful cohabitation, while accepting the many differences and conflicting interests, would bring about an economic and cultural flowering in the region.

During my earliest years in Israel—when I was a young adult—there was always a dismissiveness toward the Arab people. It was a kind of ignorant, unthinking racism. If someone did shoddy work, it would be dismissed as *avoda aravit*, or "Arab work." Certainly Ashkenazi Israelis, with their European heritage steeped in Western culture, exhibited strong prejudices about the Levant and its people—and not only about Arab Muslims but also Arab Christians, not to mention Sephardic Jews. Today, many decades later, this attitude of colonial superiority has been supplanted among some in Israel by attitudes even uglier.

At its most extreme, the taking of the life of a Jew is seen as murder, while the taking of an Arab life enjoys religious sanction. This stance was dramatized in the Israeli television miniseries *Our Boys*, based on the true story of an Arab youth who was kidnapped and murdered in retaliation for the killing of three Israeli youths by Palestinian militants. Atrocities, of course, have occurred on all sides. Among Israeli Jews, hatred is nourished not just by the preaching of Orthodox doctrine, supported by religious law. For Mizrahi Jews—those whose families originated in Arab countries—it is also fueled by fear and resentment stemming from historical oppression, as well as by a need to put distance between themselves and peoples with whom they actually have much in common. The Israeli right, many of whom are Mizrahi Jews, is now the dominant political force in Israel.

Meanwhile, attitudes toward Israel—and toward Jews—overseas are often poisonous. A few years ago, Michal and I were at a small, private dinner in England—at the country residence of a prominent architect—among a left-wing, enlightened, intellectual crowd. Yet, as the subject turned to the Middle East, the air could be cut with the resentment toward Israel, confusingly blended with anti-Semitic overtones. To be at dinner as an Israeli is to be onstage and in the witness stand. I have spent much of my life in that witness stand. The experience is only getting worse. There are moments when I've wanted to allow myself to be the critic of the government's actions, as we do while in Israel on Friday evening soirees, after Shabbat dinner, when Israelis vent their frustrations and criticisms. But this is impossible in a belligerent setting.

Being in the state of mind of mediator means constantly searching for a resolution. I have not lost my optimism that some form of peaceful cohabitation will eventually emerge, tattered though my optimism may be. Whatever comes to pass, I do not anticipate some grand bargain—some sequel to Camp David, hands joined in resolve—but rather a result that is the fruit of countless small actions pursued by countless people over the course of many years. The rich and diverse culture of the Middle East as a whole, which I cherish, has become an ineradicable part of Israel itself. It manifests itself in literature, in cinema, in music, in cuisine. The current Hebrew language is laden with Arab words, just as the speech of Israeli Arabs is laden with Hebrew words. Decades ago, subtleties in dress, haircuts, and accents made it easy to tell an Arab from a Jew. Today, this is no longer the case. Tourists to Israel are often confused and unable to tell one member of the conflict from another.

In my quest to reach out, I seek every opportunity to pursue projects and other professional ties with Arab countries. I vividly recall a moment in Saudi Arabia, in 2014. I and two colleagues were visiting in connection with work on a major urban project for a Saudi client—the mixed-use community known as Mayasem, in north Jeddah. As it happens, we were in the city during Ramadan. After midnight,

as fasts were broken and Jeddah came alive, we strolled with
our hosts down to the Old City. Suddenly, I was approached
by three young men wearing traditional dress and kaffiyehs.
"Mr. Safadi!?" they said, emphasizing the common Arabic pro-
nunciation of my name. I turned in surprise. They introduced
themselves as students of architecture at the University of
Jeddah. They admired my work. They wanted a photograph
together. I was moved, in a way that transcended what I
would have felt under similar circumstances in Shanghai or
Stockholm. Architecture had created a bond that was, you
might say, "across enemy lines."

* * *

At its best, architecture has a demonstrated capacity
to rise above the fissures that divide us. This may be
especially true when architecture aspires to capture humanity's spiritual aspirations. In 2007, while visiting Singapore,
I got a call from an Israeli friend advising that he had arranged a meeting for me with Nursultan Nazarbayev, the
president of Kazakhstan. Apparently, there were a number
of prospective projects to be discussed. The president was
vacationing in Dubai, so it was suggested I fly there to meet
him. My friend's partner, an influential Turkish hotelier and
developer named Fettah Tamince, would be present at the
meeting. When I landed in Dubai, I received a message from
Fettah saying that the president had fallen ill and that he and
his party had unexpectedly flown back to Kazakhstan. The
message added, "I've arranged a meeting for you with Sultan
Ahmed bin Sulayem, the chairman of Nakheel." Nakheel is
Dubai's largest development company. Everyone has seen
pictures of its signature project, Palm Jumeirah, an artificial
archipelago jutting out into the Persian Gulf in the shape
of a palm tree.

I was picked up at the hotel and taken to the Nakheel
offices, where I met bin Sulayem and his staff, including a
Lebanese-trained Palestinian architect, Rula Sadik. After
showing me a model of Palm Jumeirah, with its many components then in various stages of construction, bin Sulayem
indicated the entrance, where a bridge from the mainland
gave access to the island. He said, "We're going to build a
grand mosque here, and I would like you to design it." He

A detailed
sectional
rendering of the
Palm Jumeirah
Gateway Mosque,
Dubai, 2008.

followed up quickly with some details—it would have to accommodate two thousand people; it would serve as the iconic gateway to the Palm—but I was still recovering from the surprise. A mosque designed by an Israeli?

There was a moment when bin Sulayem left the room, and I was alone with Rula Sadik. I turned to her and said, "Does he know what he's doing?" I may also have said, "Is he out of his mind?" Everyone knew that I was Israeli, and it was not hard to see that this fact might entail certain complexities in a Muslim emirate, even one as tightly controlled by a ruling family as Dubai has been. When I returned to Boston, I sent a memorandum. I titled it "Full Disclosure," and I described my office and activities in Jerusalem and explained that I was an Israeli citizen as well as a Canadian and American citizen. I reiterated that I would be honored to take on the project but wanted the client to fully appreciate the context. I don't know if I was expecting some subtle reversal by way of reply, but that was not what I got. The commission was reaffirmed.

This was a task to be treated as a calling, a task that is more than just coming up with a design but rather strives to embody what connects us all. I knew something about Islam and its rituals, but I certainly was no authority. I felt the need to immerse myself deeply in the subject and, among other things, to learn about the history and evolution of mosques. One name immediately came to mind—Oleg Grabar, emeritus professor of Islamic art and architecture at Harvard, whom I knew from my university days. Grabar had retired and was now established at the Institute for Advanced Study, in Princeton. I asked him to be my consultant on the project, to inform me as well as to review the design as it proceeded. Happily, he agreed. An elegant, French-born Turk from Istanbul, Grabar had written many books on Islamic art and architecture. We began by reviewing the history of mosque building in various traditions—Arabian, Persian, Ottoman. Grabar gave me an intensive reading list. We discussed the differences among forms of Islam. We explored the spiritual meaning of a mosque's architectural features.

As I set out to work, sketching in my notebooks, Grabar was in effect looking over my shoulder. I began with a hollow sphere, which I "floated" on pylons above a pool. The highways entering the Palm archipelago passed underneath. The sphere hovered about five feet above the surface of the pool. Inside the sphere, I set the floor level of the mosque one-third of the way up. I then sliced away part of the sphere, on a slant, turning that part into a large, concave surface, like a radio telescope facing the sky, and giving the sphere the profile of a crescent moon. In the cavity that was formed, I placed a dome, suspended within the concave surface. The perimeter was perforated with skylights. The topmost portion of the sphere became the minaret of the mosque, reaching up toward the sky; there would be no separate tower. Under the sphere, between the pylons, would be the place of ablution, or purification. Water would pour from the sphere's bottom and into the ablution pool. The praying public would then rise by grand stairs and elevators to the sanctuary above.

The mosque, hovering above the water, as expressed in a model.

Grabar was encouraging. He showed me many precedents where minaret and mosque were integrated into a single structure. He felt that the allusion to the moon and its cycles, and the lighting of the mihrab—the "altar" toward which Muslims face when they pray—by skylights resonated with the traditions and symbols of Islam. Yet all this was reinterpreted in a contemporary structure that would require advanced engineering and technology to realize. Soon I was back in Dubai with models and renderings, and I experienced the pleasure, rare as it is, of jaw-dropping satisfaction on the part of the client. Some days later, bin Sulayem presented the design to Sheikh Mohammed bin Rashid Al Maktoum, Dubai's ruler, and I was encouraged to develop the plans further for structural design. I once again engaged Buro Happold, with whom we have collaborated on many challenging projects. Estimates and construction documents were prepared.

And then came the 2008 financial crash. Dubai was in chaos. There were reports of people abandoning homes and leaving luxury cars in the airport garage as they departed the country. The population of Dubai was made up primarily of expatriates, and when the economy sank, they were off. Construction projects totaling almost $600 billion were immediately canceled or put on hold.

Nakheel took a very hard hit, and the mosque project was suspended and remains so. The disappointment I felt was deep. I have always held out hope that the design will eventually see the light of day. In 2019, I met with bin Sulayem. Relations between Israel and the Persian Gulf States, which once had to be conducted at the level of an open secret, are today formalized. Bin Sulayem was in fact in Jerusalem when we saw one another. He smiled, recalling our days working on the mosque, and said, "I will get it built one day.

* * *

I have often wondered about the connection between religiosity and spirituality. I am a Jew but not an observant one. I do not follow the ritual rules and laws, though I know myself to be a spiritual person. I do believe that spirituality, or the quest for it, is a universal quality, imprinted into the makeup of our species and independent of any particular view of God or any set of practices. At the time when he was designing Notre-Dame du Haut, in Ronchamp, France— one of the great churches of modern times—Le Corbusier explained that, though he was a nonbeliever, he nevertheless felt able to design an inspiring religious building. His point was that spirituality is not the monopoly of orthodox practitioners, and that there are ways of spirituality that transcend organized religion. Louis Kahn expressed similar sentiments. By identity he was deeply Jewish, and he was also a nonbeliever. Yet he conceived the Mikveh Israel Synagogue, in Philadelphia—a compelling design, even if never built. He designed the First Unitarian Church in Rochester, New York, with its ingenious natural lighting, and a mosque in Dhaka, Bangladesh, a series of interpenetrating cylinders, which forms part of his famous parliamentary complex.

I identify with Kahn and Le Corbusier in their pursuit of a "spiritual" essence that all people have in common. I am fascinated if not obsessed by the prospect of creating places of spiritual uplift. I have always been eager to explore opportunities for engaging not only with spirituality but also with places of worship. In 2019, we received a call from Miral, the company in charge of building Abu Dhabi's major cultural institutions, to take part in an invited competition for the design of something called the Abrahamic Family House. This was envisioned as a complex that would incorporate a mosque, a church, and a synagogue; it would be located

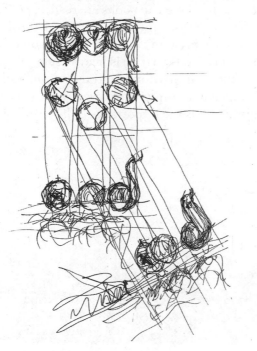

Notebook sketches for the Abrahamic Family House, 2019.

in Abu Dhabi's museum precinct, overlooking Jean Nouvel's Louvre Abu Dhabi, Norman Foster's Zayed National Museum, and Frank Gehry's Guggenheim Abu Dhabi. Once again, I confronted my own disbelief when the client acknowledged that my Israeli and Jewish background was clearly understood and would not be an issue.

The Abu Dhabi project was something new—juxtaposing, side by side, places of worship for the three Abrahamic religions. Not only would each place of worship need to capture the essential spirit of its own tradition, but the three structures would have to be in dialogue with each other, offering the hope that, in terms of architecture at any rate, relations among the religions might become an occasion for reconciliation and peace.

To design a church, a synagogue, and a mosque, I had to return to the study of the rituals and symbols of the three great Abrahamic traditions. They have much in common and much that is different. The ritual of the mosque includes the process of ablution; communal praying, with all kneeling toward Mecca; and the tradition of the call to prayer, with a spire and muezzin tower—the minaret. The synagogue is more of a community hall. The praying is toward the center, the bimah, where the books are read, as well as toward the ark, facing Jerusalem, where the Torah scrolls are kept. For its part, a Christian church is focused on the altar and the celebrant, and the ritual is heavily supported by visual and sculptural narrative and representation, both of which are forbidden in mosques and synagogues. Common to all three traditions is the importance of light and of the sheer

The Abrahamic Family House in model form, 2019.

evocative quality of organized space. One does not need to be a Muslim to experience the sensual uplift of the ancient blue-tiled mosque in Isfahan, Iran. One does not need to be a Catholic to be awed, spiritually, upon entering a cathedral such as Chartres: the soaring structure, the penetration of light, the narrative unfolding with the aid of stained-glass windows and statues and paintings. In the end, the architecture must come to the service of achieving spirituality. In the Gothic cathedral, the interplay of light and structure and geometry makes one's heart beat in a special way. The addition of music only amplifies the experience.

Several workshops were set up with the client to discuss the development of the design. The client's team was headed by Mohamed Abdalla Al Zaabi. On our first trip, after a visit to the Louvre Abu Dhabi, one of Nouvel's greatest works, we proceeded to present our drawings. We drew heavily on the unbuilt mosque design for Dubai and the synagogue at Yeshiva Porat Yosef in Jerusalem—and this turned out to be a mistake. I don't recall the exact words, but the reaction to the mosque design was along the lines of: "This is too much like your mosque in Dubai. We know that design and have studied it carefully. We love it—this is why we invited you. Yet we believe this complex demands a unique solution, a particular response to the spirit of the Abrahamic religions cohabiting in one place, in Abu Dhabi."

We needed to start from scratch. The flight home to Boston was not the happiest one. But sometimes a setback can provide a jump start. Sketching on the plane, I saw a way in which three spherical chambers might evolve as variations out of a singular encompassing form. I envisioned delicate, trellis-like structures, whose softening, permeable engagement with outside light would create caressing spaces inside; at night, the structures would emit light from within.

Each of the three spherical chambers evolved differently. The mosque generated an upward-reaching leaflike structure, to some recalling a cobra. It would both serve as the minaret and direct light on the mihrab below. For the synagogue, I followed the Sephardic ritual tradition with which I grew up: sitting in the round as a community. Hence, within the sphere, a second toroid-like sphere descended, hovering over the bimah. The tension in the synagogue between the community (focused on the bimah) and the deity (focused on the *aron hakodesh*, the chamber that contains the Torah scrolls), with its orientation toward Jerusalem—was dramatically expressed in the architecture. For the church, the sphere was split into two spheres of different radii, which created a cavity in which church bells could be visibly suspended.

For the altar design, in the church, I drew on an unforgettable experience I had had with Jorge Enrique Jiménez Carvajal, the archbishop of Cartagena, Colombia. In designing a church, round in form, for the community there, I had proposed an altar in which the curved concrete wall had carved through it a large cross. What made this cross different was that it leaned sideways, 30 degrees off the vertical. It was a dramatic departure from the traditional upright cross. The archbishop asked me about the leaning cross, and I explained that I have always been emotionally drawn to images of Jesus carrying the cross, where it weighs on his shoulders at an angle; the humility and suffering are palpable. The archbishop smiled. At the end of a long conversation, he said, "Perhaps we should decree this for all churches in the future."

In our design for Abu Dhabi, we proposed that the three spheres be placed over a pond, flooding the entire site as a reflecting pool. One could descend into the pond's perimeter, as in a Hindu water temple. The three spheres would hover above the water and also be reflected in the water—seemingly weightless, defying gravity. In the space nestled between the three spheres, at the center, we located the Abrahamic visitors' center.

Elements of the Abrahamic Family House: the mosque.

The Abrahamic
Family House:
the church.

Months went by without news of a decision. I wasn't
entirely surprised. There had been intense conflicts in the
Gulf area among the United Arab Emirates, Iran, and Qatar;
the sheikh was preoccupied. Judging from the reaction of the
CEO and staff, we thought the commission would be ours.
Alas, we finally learned that a design by the Ghanaian-British
architect David Adjaye had been selected. Despite my deep
disappointment, I still believed that our design was one of
the most significant and enduring of my lifetime. I had been
able to express themes and thoughts that had been yearning
for expression for a long time. I know that the ideas and
concepts embodied in this design will ultimately be realized—
somewhere, someday.

* * *

My experiences as an architect with several countries in
the Middle East, with issues of peace and with expressions of the human spirit, have not occurred in a vacuum.
One way or another, they reflect the all-too-present reality
of life on the ground. Over the years, my thoughts about the
resolution of the Israeli-Palestinian conflict have changed—
just as actual conditions have changed. There are urgent new
problems and perhaps new opportunities. The boycott of
Israel by the entire Arab and Muslim world has loosened
during the past few years. At the same time, the Palestinian
refugee issue remains unresolved. The Arab Spring has come
and gone. Terrorism remains a daily threat. We have moved
from a world dependent on Middle Eastern oil to one with ·

With client Daniel
Haime (far right),
and colleagues
Sean Scensor
(standing, right),
and Paul Gross
(standing, center)
as we review the
design of a church
with Jorge Enrique
Jiménez Carvajal,
then archbishop
of Cartagena,
Colombia.

diminishing dependence on it. Israeli and Palestinian societies have evolved and transformed. Israeli-Palestinian peace negotiations are a priority for virtually no one.

In my naive youth, I had experimented with designs for cities for Palestinian refugees on the West Bank and in Egypt, implicitly accepting the concept of benevolent occupation. After 1967, like many Israelis, I came to believe in the two-state solution as the way to resolve both the conflict and the occupation, by establishing a Palestinian state on the West Bank and in Gaza. After the Oslo Accords, in 1993 and 1995, and up until the Rabin assassination, a two-state solution seemed plausible. Now, two decades of right-wing governments in Israel have shattered the Oslo Accords, deepened the occupation, and expanded the settlements. The Palestinians are divided among themselves and segregated into disconnected pockets. Gaza, under Israeli blockade, is controlled by Hamas.

These developments, in combination, have shifted the debate. Some people still strive for a two-state solution. Others now look to a one-state solution in some form— that is, Israelis and Palestinians in a single entity. Each of these concepts has supporters and detractors on both the left and the right, and among both Israelis and Palestinians. The underlying problem that I believe makes both proposals unworkable is the extraordinary asymmetry of power, technology, and financial clout between Israel and Palestine. This has led me to consider another option, one that might combine the positive elements of both approaches: the idea of a single federated state, one in which there would be common administration in the domains of security, resources, transportation, and foreign affairs, but where two national communities would have individual control over many "domestic" responsibilities such as education, social services, immigration, and certain aspects of economic life. I am not under any illusion that this idea, which I spelled out some years ago in more detail and circulated to people I knew in Israel and the United States, is a blueprint that all parties will suddenly unite behind. I do believe it is the responsibility of people of good will to chip away at the problem relentlessly. Sometimes more progress can be made than anticipated, and bridges may be built in unexpected ways. Through the act of building itself, I have sought to do my own chipping away.

A few years ago, I was invited by the BBC to participate in a live public interview as part of a series about architects and their work and beliefs. There were many Palestinians in the audience, and several asked me probing questions about my activities in Israel. I answered candidly.

The most surprising question was the last one. Given my experience designing Yad Vashem, would I be willing to design a Palestinian museum commemorating the Nakba? The Arabic word *nakba* means "disaster" or "catastrophe" and refers to the war of 1948, when hundreds of thousands of Palestinians fled or were expelled from their homes in what would become Israel. My response was that I believed that a project of such national significance to the Palestinians should be designed by a Palestinian architect. The follow-up question was this: if invited to collaborate with a Palestinian architect in designing a Nakba museum, would I accept the invitation? I responded that I would be deeply honored.

Interior of the
Class of 1959
Chapel, Harvard
Business School,
Cambridge,
Massachusetts.

What I Believe

There is mystery at the heart of architecture just as there is mystery in the meaning of life. Life and architecture are bound together, and the principles that guide one as an architect cannot be separated from the principles that guide one as a human being. Nor can they be separated from the principles that guide communal life as we grope toward justice and beauty.

There are some who do not agree with this. I think of Philip Johnson's battle cry, cited earlier: "There is only one absolute today and that is change. There are no rules, surely no certainties, in any of the arts. There is only a wonderful freedom." We see the results all around us: those private jokes in public places. But to think this way—to build this way—is to misunderstand the responsibility of an architect. Indeed, it is to betray that responsibility.

Architecture is not a freewheeling expressive art. The expressive power of architecture is not an end in itself. Architects must serve society by creating the physical setting for life's activities. In the past—during the Middle Ages and the Renaissance, for instance—the passion stirred by Christian belief led to the creation of great works of art in the realms of painting, sculpture, music, and architecture. We cherish those achievements, but the bulk of the population lived in a condition of utter misery. With the emergence of the modern movement, at the turn of the twentieth century—one of the most remarkable moments in the history of architecture—the aristocratic and elitist framework, beholden to the demands of princes and priests, of aristocrats and plutocrats, at last had competition. Manifestos by architects proclaimed a responsibility to society as a whole. True, the manifestos often had an elitist framework of their own; the focus was often political. But the motivations went deeper, to a new notion of individual human rights and the needs of an entire society. These motivations inform my work.

* * *

I do not believe in a godly being dedicated to governing and controlling us, rewarding and punishing us. I accept how much we do not know about the universe or about our place in it. There are secrets we may never unlock. As a

species inhabiting the planet—with a measure of conscious-
ness and intelligence that we believe exceeds those of other
organic beings known to us—we must figure out how best to
proceed without the benefit of prescription from above. I
seek to derive ethical and moral principles ultimately from
compassion. For some, compassion may come across as an
abstract ideal, a disembodied virtue, but like altruism—maybe
like all the great virtues—it is deeply pragmatic. In the long
run, and even the short run, my own interests, and those
of my family and of others I love, are entwined with the
interests of every other fellow being, human and otherwise.

The principles that guide me as a person precede the
principles that guide me as an architect. They have a profound impact on the process of evolving a design—whether
for housing or a library or a commercial structure or a shrine.
I am unable to design in a manner that considers the benefit of the few at the expense of the many; unable to design
high-quality spaces for some and inferior spaces for others.
There will always be inequities, but there must be a reasonable balance. The impact on architecture and urbanism of
the quest for balance is profound.

The quest for reasonable balance leads to a series of
prescriptions—rules of engagement for my life as an architect.
In talking them out, I would begin by noting that architecture
is in the service of people, and therefore must respond to the
life and activity intended to occur within a building, be it a
home, a school, an office, a museum, a resort, or a place of
worship. There is a purpose greater than the structure itself,
and architecture must accommodate that purpose and reso-
nate with it. Further, we need to always be responsible with
respect to resources, building methodologies, and energy.

A building must "fit" and "belong" and interact with its
immediate site—the topography, the climate, the surrounding
structures, and in a broader sense the history and culture
of a particular place. The marching orders for an architect
come in the form of what is called the program—the list of
spaces and facilities required for the type of building it is.
The program and the site together allow an architect to begin
developing a plan. Le Corbusier used to say, "The plan is the
generator." In truth, the site itself is probably the generator;
the plan is the outcome of a merging of program and site. A
site possesses a set of properties, some of them obvious and

some of them elusive. Every site harbors secrets. Every site—a slope overlooking the mountains, a flat plain on a riverbank, a street corner giving onto a square—offers opportunities. Geographic location tells us more. The path of the sun from sunrise to sunset, from winter solstice to summer solstice; the seasons; the rains; the prevailing winds: all of these are considerations that architecture should respond to. Given the extraordinary range of climates, it is nonsensical that we design almost identical buildings around the world.

In laying out the rules of engagement, we must acknowledge that many of the quality-of-life amenities we have traditionally expected a building to provide become great challenges in the modern, dense, congested high-rise city. Simple things like privacy, identity, and access to light and nature become exponentially more difficult to achieve as density increases. A world of 10 billion inhabitants presents unprecedented challenges. How can architecture promote individual well-being and a balance with nature in the age of megascale?

* * *

Architecture's medium is fundamentally material. In the not-too-distant past, the material palette was limited to wood, brick, and stone, to mud and straw. At some point in the even less-distant past, we added concrete, steel, and glass. This limited palette made life relatively simple for architects: familiar materials could be deployed in familiar ways. Not so today, with a wide variety of materials, textures, and methods to choose from, and therefore the possibility of endless choices. The more choices there are, the greater the temptation to self-indulgence—and the greater the need for responsibility. What once was a truism—that vernacular architecture evolved incrementally, as in nature, from one small mutation to another, toward some level of greater adaptability and fitness—is hardly true today. The culture urges invention. It seeks newness for its own sake.

While the postmodernist movement, with its emphasis purely on form, on the merely stylistic, has passed by, it has left a legacy of permissiveness that plagues the profession to this day. What was once a socially conscious profession—one with high ideals and a sense of mission, akin perhaps to those of medicine—is shaped by the forces of commercialism and branding, of fashion and faddishness. It is deeply influenced by the values and machinations of the art and fashion worlds. These in turn are driven by the market, a market that "knows best" and in turn drives the actions of developers.

The expressive power of an architect—what she or he can make a physical structure look like—need not come at the expense of a higher mission. Architects must develop designs that resolve the needs of everyone. They must use resources responsibly. They must be guided by empathy and compassion. As I meet and work with architects who are early in their careers, one characteristic I have encountered in recent years is a reawakening to social issues and the basic needs of society as a whole—the welfare of humanity, the welfare of the planet. I hope that this will manifest itself in their work. I needed time before understanding how to embody it in my own.

In the end, architecture is not purely subjective—it is not simply about aesthetics or taste. There are rights and wrongs. How all of this comes together—the moral and social compass, the objective analysis, the progressive iteration, the intuition, the integration—is the mystery of architecture, and again, it is perhaps not unlike the mystery of life itself.

Once, four decades ago, in Jerusalem, I set out to distill my beliefs into a kind of poem. It went like this, and I hold to it still:

He who seeks truth shall find beauty.

He who seeks beauty shall find vanity.

He who seeks order shall find gratification.

He who seeks gratification shall be disappointed.

He who considers himself the servant of his fellow beings shall find the joy of self-expression.

He who seeks self-expression shall fall into the pit of arrogance.

Arrogance is incompatible with nature.

Through nature, the nature of the universe and the nature of man, we shall seek truth.

If we seek truth, we shall find beauty.

Acknowledgments

This memoir is primarily a chronicle of my life as an architect. Although by necessity it touches on many personal details, its focus is on the making of architecture. This has been an all-consuming endeavor, one that has had an impact on the lives of my family.

I am deeply indebted to my wife of forty-one years (and my partner of fifty years), Michal Ronnen Safdie, my lover, my friend, my trusted counselor. She has lived through many of the adventures described here—a constant adviser who has taught me to be more attentive and appreciative of those around me. She witnessed the writing of this memoir, read each draft, corrected my memory, issued the occasional warning, and wielded a discerning pen. This book is dedicated to her.

My children—Taal, Oren, Carmelle, and Yasmin—have had to put up with, at times, an absent father. (And the next generation—Ariel, Rafael, Raquel, Mia, Gene, and Leila—with a peripatetic grandfather.) I hope they were somewhat compensated by the travel and opportunities that came with my work. I am deeply grateful for their support and patience. All four of my children have lived with architecture all of their lives, and I have been the beneficiary of their insights.

I am grateful, too, for the love and support over many decades of the broader Safdie family. I and my brother, Gabriel, and my sisters, Sylvia and Lillian, shared two remarkable parents, Leon and Rachel Safdie. Lillian was born in Montreal, but with Gabriel and Sylvia, and of course with my parents, I also shared the wrenching experience of emigration from Israel and the challenge of starting a new life in Canada. All of this was formative.

I deeply appreciate the enthusiasm and critical engagement of my friends Stephen Greenblatt, Joseph Koerner, Homi and Jackie Bhabha, and Samantha Power. All of them read the manuscript and provided valuable comments. They, together with Ramie Targoff, Meg Koerner, and Cass Sunstein, and their children, have long been intertwined with the lives of my own family. More recently my friend Noah Feldman provided helpful suggestions. Finally, when we visited not long ago with our friends Paul and Maria Audi, in Patmos, Paul read a chapter a day and provided enlightening insights.

This memoir spans the eighty-four years of my life and particularly the sixty years of my career as an architect. The many projects realized on five continents over this period could not have been achieved without the dedication and talent of the professional and other staff in my firm, first in Montreal and then in Boston, along with those in our offices in Israel, Singapore, and China. I have worked with some colleagues for decades. I think of the members of this staff, past and present, as part of a large and extended family. There are too many to name individually; some are mentioned in the book itself. This is a moment to salute them all. Special thanks go to Natalie Wanjek and Robyn Payne, whose research help was essential in the making of this book.

Several times over the past decade, I had set out to write a memoir. Time and again, I was thwarted by the pressure of various projects and the travel the projects entailed. When travel was suspended owing to the onslaught of the Covid-19 pandemic, I was encouraged to make yet another start by Samantha Power. A dear friend of Michal's and mine for twenty-five years, Samantha went beyond encouragement and suggested I get an editor to assist me in organizing the material and helping to craft the narrative. Happily, she introduced me to Cullen Murphy, editor at large at *The Atlantic*. I am deeply indebted to Cullen for helping to draw out events and ideas, and even grace the text with humor. In the process, I have gained a friend. Cullen introduced me to Raphael Sagalyn, who became my literary agent and in turn introduced me to Grove Atlantic and my editor, George Gibson. George was a constant reminder of the broader audience—the person who doesn't know much about architecture but wants to understand more. Thanks to him and the entire professional team at Grove Atlantic. Thanks also to Michael Gericke and his colleagues at Pentagram for the design of this book, and for their collaboration on many other books and projects over a period of several decades.

At eighty-four, I feel blessed—blessed by family, friends, and colleagues; by endless opportunities; and by the ability to pursue the work I love, which I intend to continue as long as my mind and body allow.

Aleppo, 1916: the Safdie clan at a family wedding. My father is second from the right in the top row.

With my mother, Rachel, c. 1942.

My brother, Gabriel (left), and sister Sylvia (center), 1952.

My sister Lillian in the arms of my father on the day I graduated from McGill, 1961. My mother is at left, my first wife, Nina, at center.

With Nina and our children, Taal and Oren, 1969.

With my father in Jerusalem, 1984.

My wife, Lieutenant Michal Ronnen, not long before we first met, in 1972.

With the extended Safdie family in Mexico, celebrating the eightieth birthday of my mother (with hand raised), 1994.

Michal's and my daughter, Yasmin (right), with her wife, Deb, and their daughter, Leila.

My daughter Taal (right) and her husband, Ricardo (left), with their children, Ariel, Rafael, and Raquel.

Michal Ronnen Safdie at our home in Cambridge.

My son, Oren with his wife, M.J., and their daughter, Mia.

Michal's and my daughter Carmelle with her husband, Spencer, and their son, Gene.

Celebrating my eightieth birthday with children and grandchildren, 2018.

Rendering of the proposed Osaka Integrated Resort, Osaka, Japan, 2019. The complex would have been built on an artificial island in Osaka.

Projects by Moshe Safdie

Significant Built

1967	Habitat '67	Residential	Montreal, Quebec, Canada
1970	Yeshiva Porat Yosef	Academic	Jerusalem, Israel
1981	Callahan Residence	Residential	Birmingham, Alabama
1981	Coldspring New Town	Residential	Baltimore, Maryland
1985	Ardmore Habitat Condominiums	Residential	Singapore
1987	Quebec Musée de la Civilisation	Museum	Quebec City, Quebec, Canada
1988	Cambridge Center	Mixed Use	Cambridge, Massachusetts
1988	National Gallery of Canada	Museum	Ottawa, Ontario, Canada
1989	New City of Modi'in	Urban Design/Planning	Modi'in, Israel
1989	The Esplanade	Residential	Cambridge, Massachusetts
1991	Montreal Museum of Fine Arts	Museum	Montreal, Quebec, Canada
1992	Harvard Business School Master Plan, Morgan Hall, and Class of 1959 Chapel	Academic	Boston, Massachusetts
1994	Rosovsky Hall, Harvard-Radcliffe Hillel	Academic	Cambridge, Massachusetts
1994	Ottawa City Hall	Government	Ottawa, Ontario, Canada
1995	Ford Centre for the Performing Arts	Cultural	Vancouver, British Columbia, Canada
1995	Vancouver Library Square	Library	Vancouver, British Columbia, Canada
1995	Neve Ofer Community Center	Mixed Use	Tel Aviv, Israel
1998	Hebrew Union College	Academic	Jerusalem, Israel
2000	Exploration Place	Museum	Wichita, Kansas
2002	Hebrew College	Academic	Newton, Massachusetts

Significant Built (continued)

2003	Cairnhill Road Condominiums	Residential	Singapore
2003	Salt Lake City Public Library	Library	Salt Lake City, Utah
2003	Peabody Essex Museum	Museum	Salem, Massachusetts
2003	Eleanor Roosevelt College	Academic	La Jolla, California
2004	Ben Gurion International Airport, Airside Terminal	Airport	Tel Aviv, Israel
2005	Yad Vashem Holocaust History Museum, Children's and Transport Memorials	Museum	Jerusalem, Israel
2006	Jepson Center for the Arts, Telfair Museums	Museum	Savannah, Georgia
2007	Lester B. Pearson International Airport, Terminal 1	Airport	Toronto, Ontario, Canada
2008	Bureau of Alcohol, Tobacco, Firearms and Explosives	Government	Washington, D.C.
2008	United States Federal Courthouse	Government	Springfield, Massachusetts
2009	Mamilla Center, Citadel Hotel, and David's Village	Mixed Use	Jerusalem, Israel
2010	Yitzhak Rabin Center	Cultural	Tel Aviv, Israel
2011	Marina Bay Sands Integrated Resort	Mixed Use	Singapore
2011	Khalsa Heritage Centre	Museum	Anandpur Sahib, Punjab, India
2011	Crystal Bridges Museum of American Art	Museum	Bentonville, Arkansas
2011	Kauffman Center for the Performing Arts	Cultural	Kansas City, Missouri
2011	United States Institute of Peace	Government	Washington, D.C.
2013	Skirball Cultural Center	Cultural	Los Angeles, California
2016	Sky Habitat	Residential	Singapore
2017	Eling Residences	Residential	Chongqing, China
2017	Habitat Qinhuangdao	Residential	Qinhuangdao, China
2018	Jewel Changi Airport	Mixed Use	Singapore
2019	Free Library of Philadelphia	Library	Philadelphia, Pennsylvania
2019	Monde Residential Development	Residential	Toronto, Ontario, Canada
2020	Raffles City Chongqing	Mixed Use	Chongqing, China
2021	Altair Residences	Residential	Colombo, Sri Lanka
2021	Albert Einstein Education and Research Center	Health	São Paulo, Brazil

2021	Serena del Mar Hospital and Los Morros Master Plan	Health, Urban Design/ Planning	Cartagena, Colombia
2021	Qorner Tower Apartment Complex	Residential	Quito, Ecuador
2022	LuOne	Mixed Use	Shanghai, China
2022	Orchard Boulevard, The EDITION Singapore	Residential/Hotel	Singapore
2022	Surbana Jurong Campus Headquarters	Office	Singapore
2022	National Campus for the Archaeology of Israel	Research	Jerusalem, Israel

On the Boards (as of May 2022)

Crystal Bridges Expansion	Cultural	Bentonville, Arkansas
Tokyo Integrated Resort	Mixed Use/ Integrated Resort	Tokyo, Japan
Marina Bay Sands Expansion	Mixed Use/ Integrated Resort	Singapore
Habitat Qinhuangdao, Retail and Phase 2	Residential	Qinhuangdao, China
Shenzhen Airport, Project Crystal	Airport	Shenzhen, China
Asia House	Mixed Use	Tel Aviv, Israel
Meta/Facebook, Project Uplift	Office/Mixed Use	Menlo Park, California

Illustration Credits

For any image not listed here, every effort has been made to trace the copyright holders and obtain their permission. The publisher apologizes for any errors or omissions and would be grateful if notified of any corrections that should be incorporated in future editions of this book.

0	courtesy Safdie Architects	58–59	Louis I Kahn Collection, the University of Pennsylvania and the Pennsylvania Historical and Museum Commission
2	Darren Soh		
3	courtesy Safdie Architects	61	courtesy Safdie Architects
4	Newsweek	62	Sam Tata
5	courtesy Safdie Architects	64	Ian Lemko
6	© 2021 The LEGO Group	65	courtesy Safdie Architects
10	CapitaLand	68	Archives de la Ville de Montréal, VM94-EX36-003
11	Shao Feng		
14	courtesy Safdie Architects	69	Arnott Rogers
16	Sueddeutsche Zeitung Photo (Alamy)	74–75	courtesy Safdie Architects
19	Nazareth Palestinian Archive	76	(bottom) © Estonian Museum of Architecture by the gift of Merike Komendant Phillips and G. Jüri Komendant
20	Palphot (postcard)		
22	courtesy Safdie Architects		
23	from the Collection of Studebaker National Museum, South Bend, Indiana	77–79	courtesy Safdie Architects
27–35	courtesy Safdie Architects	80	Peter Porges / the New Yorker Collection / the Cartoon Bank
36	Hemis (Alamy)	81	Jerry Spearman
38	courtesy Safdie Architects	82–83	courtesy Safdie Architects
44	Penguin Random House	85	Newsweek
47	© F.L.C. / ADAGP, Paris / Artists Rights Society (ARS), New York 2021	86	Niday Picture Library (Alamy)
49	mccool (Alamy)	91	National Photo Collection of Israel (Public Domain)
50	Ralf-Finn Hestoft (Getty Images)	92	Michal Ronnen Safdie
51–53	courtesy Safdie Architects	93–95	courtesy Safdie Architects
55	Malcom Smith	96	Jerry Spearman
56	Photo Researchers (Getty Images)	97–98	courtesy Safdie Architects
57	Louis I Kahn Collection, the Architectural Archives, University of Pennsylvania	106	(top) Public Domain

Insert Illustration Credits

Index

Bold numbers denote references to illustrations or photographs.